Springer Polar Sciences

This Series is now being indexed by SCOPUS.

Springer Polar Sciences is an interdisciplinary book series that is dedicated to research in the Arctic, sub-Arctic regions, and the Antarctic. In recent years, the polar regions have received increased scientific and public interest. Both the Arctic and Antarctic have been recognized as key regions in the regulation of the global climate, and polar ecosystems have been identified to be particularly susceptible to the ongoing environmental changes. These changes are having widespread implications for human communities, businesses, and governance systems and are interacting with demographic shifts, globalisation, resource development, cultural change, territorial disputes, and growing calls for self-determination in some regions. Consequently, the international efforts in polar research have been enhanced considerably, and a wealth of new findings is being produced at a growing rate by the international community of polar researchers and those who live in the region.

Springer Polar Sciences aims to present a broad platform that will include state-of-the-art research, bringing together both science, humanities, and perspectives rooted in indigenous and local knowledge to facilitate an exchange of knowledge between the various polar science communities. The series offers an outlet to publish contributions, monographs, edited works, conference proceedings, etc. Topics and perspectives will be broad and will include, but not be limited to, climate change impacts, climate change policy, environmental change, polar ecology, governance, health, economics, indigenous populations, tourism, resource extraction activities, and research design in polar regions. Books published in the series will appeal to scientists, students, polar researchers, community leaders, and policy makers.

Mathias Albert • Mikko von Bremen
Svenja Holste • Daniela Portella Sampaio
Holger Straßheim

The Worldviews of Ice

Constructions of the Arctic at the Science/Politics Interface

Springer

Mathias Albert
Faculty of Sociology
Bielefeld University
Bielefeld, Nordrhein-Westfalen, Germany

Mikko von Bremen
Faculty of Sociology
Bielefeld University
Bielefeld, Germany

Svenja Holste
Faculty of Sociology
Bielefeld University
Bielefeld, Germany

Daniela Portella Sampaio
Favo 7
International Relations Research Center
São Paulo, Brazil

Holger Straßheim
Faculty of Sociology
Bielefeld University
Bielefeld, Germany

ISSN 2510-0475 ISSN 2510-0483 (electronic)
Springer Polar Sciences
ISBN 978-3-032-25944-8 ISBN 978-3-032-25945-5 (eBook)
https://doi.org/10.1007/978-3-032-25945-5

This Springer imprint is published by the registered company Springer Nature Switzerland AG
The registered company address is: Gewerbestrasse 11, 6330 Cham, Switzerland

Acknowledgements

This book is the result of a research project of the same name funded by the German Research Foundation (DFG) from 2022 to 2025. We gratefully acknowledge DFG support for this research. Institutional support has been provided by the Institute for World Society Studies at the University of Bielefeld. Sebastian Knecht co-designed an earlier version of the project. We would like to thank Paul Reuber and Andreas Läufer for comments provided during various stages of the project, and Moritz Köster for research assistance. Stephen Curtis copy-edited the entire manuscript. Special thanks to Christoph Humrich for providing many helpful comments on the full draft.

Contents

About the Authors

Mathias Albert is Professor of Political Science at the University of Bielefeld, where he has held a chair since 2001. His research spans the history and sociology of world politics, world society theory, global conflict, polar geopolitics, and youth studies, with a particular focus on how global orders are compared, structured, and transformed over time. Albert is a member of the Institute for World Society Studies and has held numerous leadership roles in interdisciplinary research initiatives, including large-scale collaborative projects funded by the German Research Foundation. He has also served internationally, most notably as the Hannah Arendt Visiting Chair at the University of Toronto. An internationally recognised scholar, Albert has published around 30 books and more than 200 journal articles and book chapters. His major works include *A Theory of World Politics* (Cambridge University Press, 2016), which advances a sociological perspective on international relations and global order. In addition to his work on world politics, he is a leading expert in youth research and polar policy, bridging political science with sociology and interdisciplinary global studies.

Svenja Holste is a postdoc. Her dissertation at the University of Bielefeld investigated Arctic knowledge production in permafrost research and glaciology. Her research interests include international relations and science studies in polar regions, with a recent focus on questions of legitimacy and expertise, as well as co-creative research practices. Previously, she has worked as a project officer for the Association of Polar Early Career Scientists and served as co-chair of the International Arctic Science Committee's Action Group on Carbon Footprint. She is a member of the Responsible Science Initiative (RISE) and draws motivation for her work from interdisciplinary and outreach activities. She has published collaborative work on access to Arctic research infrastructures and on sustainable science practices in Arctic research.

Daniela Portella Sampaio is a research affiliate of the *Research Centre for International Relations* (NUPRI) at the University of São Paulo. Her work lies at the intersection of international relations, political theory, and public policy, with a focus on environmental, marine, and polar governance. She has held two Marie Skłodowska-Curie Postdoctoral Fellowships, most recently completing one at the Helmholtz Institute for Functional Marine Biodiversity at the University of Oldenburg, where her research examined how European governments design and implement responsibility and liability frameworks for environmental and human risks in Antarctic governance. She has extensive experience conducting applied, policy-facing research in international and transnational contexts, including work on marine protected areas, environmental risk management, and public–private governance arrangements in the Antarctic. Portella Sampaio has held research positions at Bielefeld University, the University of Leeds, and City University of London. She has worked closely with the Antarctic Treaty Secretariat for over a decade, contributing research and legal-institutional analysis to Antarctic Treaty Consultative Meetings and advising on issues of environmental protection, liability, and governance within the Antarctic Treaty System.

Holger Straßheim is Professor of Political Sociology at the University of Bielefeld. His work is dedicated to exploring public policy, governance, and knowledge in world society. He particularly focuses on expertise and democracy, the role of expertise and evidence in public policy, polar science-politics relations, the ways economic knowledge shapes social regulation, and the nexus between policy areas, such as sustainability, climate change, and global public health. Straßheim is co-director of the Institute for Science Studies (ISOS) at Bielefeld University, a member of the German National Committee for Polar Research, and a member of the Social & Human Working Group (SHWG) of the International Arctic Science Committee (IASC). He serves on the editorial board of the journal *Critical Policy Studies* and is co-editor of the *Advances in Critical Policy Analysis* book series. Among his publications are ‘A World of Evidence’ (*Policy and Society* 43, 2024), ‘Sustainability in Public Policy’ (*Encyclopedia of Public Policy*, 2024), and the *Handbook of Behavioural Change and Public Policy* (with S. Beck; Edward Elgar, 2019).

Mikko von Bremen is a doctoral researcher at the University of Bielefeld. His dissertation is on the role of Arctic research funding at the science-politics interface. His research interests also include international relations and foreign policymaking in the Arctic. In his previous work as a project coordinator at the Konrad-Adenauer-Stiftung Nordic Countries Programme in Riga and Stockholm, he wrote several reports on the politics and international relations of the Arctic region and the Nordic countries. He is a visiting fellow at the International Development and Research Network (IDRN) and contributed to a factsheet by the German Arctic Office on *Germany in Arctic Research.*

Chapter 1
Introduction: Melting Ice, Science and Politics

Abstract The introduction describes how the book seeks to provide a 'full circle' account of how knowledge about the Arctic is produced and translated at the interface between the sciences and the political system. 'Full circle' refers to how the process of *knowledge production* itself is subject to a range of factors that influence scientific research. Instead of pretending that there is some kind of 'neutral' science out there, that produces results which can be transferred to politics (and then be used there in any way whatever), the book looks at how specific worldviews that are also political in nature actually influence or even emerge out of the research process itself: firstly, in relation to how funding is allocated; secondly, in relation to the working routines of scientists; and thirdly in scientific advice.

Keywords Sciences · Knowledge production · Politics · Arctic change

The present book seeks to provide a 'full circle' account of how knowledge about the Arctic is produced and translated at the interface between the sciences and the political system. 'Full circle' means that we go beyond the many important studies that highlight how processes that transfer knowledge 'from' science 'to' politics work—and how they might be influenced by a range of social, political, cultural or other factors that could hamper the unrestricted flow of knowledge. Rather, we seek to scrutinise how the process of *knowledge production* itself is subject to a range of factors that influence scientific research. In other words: instead of pretending that there is some kind of 'neutral' science out there, that produces results which can be transferred to politics (and then be used there in any way whatever), we look at how specific worldviews that are also political in nature actually influence or even emerge out of the research process itself: firstly, in relation to how funding is allocated; secondly, in relation to the working routines of scientists; and thirdly in scientific advice.

In a somewhat idealised (if not utopic) view of the relation between science and politics, scientists would study all possible facets of the Arctic and its changing conditions, then convey their findings to bureaucrats and politicians who, on the

M. Albert et al., *The Worldviews of Ice*, Springer Polar Sciences,
https://doi.org/10.1007/978-3-032-25945-5_1

basis of sound scientific results, would deliberate proper courses of action. In such a view, politics would decide that it was important to continue funding science that provides neutral results and advice. Science, in short, would be speaking truth to power, and power would be listening.

However, the Arctic does not exist in an ideal world (nor, for that matter, does anywhere else). The things that people do are influenced by many factors; and those factors are influenced, in turn, by what people do. The purpose of this book is to arrive at a better understanding of the role that specific factors, particularly worldviews, play in what might be called an ideal-typical Arctic science/policy circle—that is the *circle* that embraces funding research, actually 'doing' the research, and then translating of the results of that research into politics (which then in turn again feeds into research funding decisions etc., continuing the circle). While these worldviews have to do with a number of different aspects, we posit that they can all be seen as 'geopolitical' in a wider sense, in that, in relation to a political field that being an international space always has an underlying geopolitical background, they pertain to who the relevant actors are or should be, how they relate to one another, which norms they adhere, or should adhere, to, and how all this translates into relations between (idealised and real) systems of science and politics. In that sense, a study like the present one has an academic purpose and can, on the one hand, be read as a contribution to areas such as the sociology of science or critical geopolitics. It could also, on the other hand, be read as a practical contribution in the sense that simply laying open the presence and roles of worldviews in the different stages of the circle contributes to an understanding of the ways in which these stages operate and are linked. It needs to be emphasised, however, that there is nothing here that even comes close to giving concrete advice. Our aim is to provide insights into the way in which worldviews come into play in the larger science/policy circle—not to provide advice on how to leave them out or alter them.

While the present study, in its empirical parts, could conceivably cover an extremely broad area (i.e. all worldviews in all Arctic research everywhere), it starts with a number of restrictions that are already aptly expressed in its title. It is called *The Worldviews of Ice* for a reason—namely, that the region's being characteristically partly or completely ice-covered and subject to ongoing changes with high degrees of ice loss provides a widely shared reference point for many more specific imaginaries—like, for example, the Arctic as a pristine wilderness, a space of indigeneity, a geopolitical rift zone, etc.[1] This is not to say that ice and ice coverage constitute an exclusive imaginary. But it is quite likely the most important one, as witnessed not least by the role of ice and snow in illustrations across a wide range of knowledge areas and depictions of Arctic issues in popular culture and the media. This special status given to 'worldviews of ice' provides us with an access point for our research—particularly when it comes to enquiring how specific imaginaries play out in what researchers do and how they do it. We concentrate on researchers who actually deal with ice 'directly', and here particularly on those in glaciology

[1] See Chap. 2 below for our take on the relation between 'worldviews' and 'imaginaries'.

and permafrost research. This concentration reflects the thematic proximity of these fields to our central narrative. It is in no way meant to claim that worldviews would be absent from 'ice-free' Arctic research, such as into questions of marine biology, geology or atmospheric physics. Within the direct dealings with ice, glaciology and permafrost research represents a fair share of all research. We opted for those, rather than research on sea ice, in particular, not because the latter tends to be most obviously connected to popular imaginaries of melting and vanishing (after all such popular imaginaries might find their way into worldviews in other areas as well). Rather, we chose to focus on what is fields which mostly do terrestrial research because such research is (to a very large part) still highly dependent on research done on national or Indigenous territories and thus most prone to also depict factors influencing worldviews that stem from references to sovereignty issues.

In looking at changing emphases in Arctic research more generally and how knowledge at the science/politics interface interlinks with a range of contexts, we focus on three countries that have strong Arctic research traditions: Canada, Germany and Norway. This reflects a variety in their characteristics both as states and research regimes that have, in terms of territory and (Indigenous) population, a very significant (Canada), a highly significant (Norway), and no (Germany) presence in the Arctic. Our underlying assumption is that variety in both knowledge regimes and Arctic politics makes a difference with regard to how worldviews are created and become influential at the science/politics interface. While our interest in these three countries serves as a focal point that allows us to concentrate our research strategy, it is worth noting that, particularly in cases of international and multilateral research cooperation, this focus is not meant to exclude on principle vistas that pertain to important players in the field from other countries. Equally important is the timespan covered. We set out to study developments, particularly trends in research funding, that at their core span the three decades from 1990 to 2020. We deliberately cover a time span that allows us to observe the long-term evolution, challenging or disappearance of worldviews of ice. However, talking to people in science and at the science/politics interface with the idea of unearthing worldviews in a broader sense, while at the same time asking them to limit themselves to a period ending in 2020, turned out to be quite unrealistic after the start of the War in Ukraine in 2022. So, particularly in Chap. 4, talking to 'real people' in the 'here and now' certainly makes a difference to what they see, even in the past.

1.1 Science in a Changing Arctic

Looking at the Arctic as a whole, one could quite provocatively state that it is a region that requires very little additional research (or, for that matter, research funding) to grasp the main lines of its development. There is little disagreement that:

- the Arctic is highly affected by global warming (indeed, apart from other regional pockets, air temperature is increasing three to four times faster there than over the rest of the globe);
- rising temperatures lead to a constant retreat of sea and land ice and snow cover, increased precipitation, the possible collapse of the Greenland ice sheet, and an increasing thawing of permafrost soil;
- these developments affect Arctic ecosystems, the living spaces of local and Indigenous populations, as well as built environments and infrastructure;
- possible side effects are associated with the political and economic interests attracted by new possibilities for natural resource exploitation and transport routes particularly those that might result from ice loss;
- while barely the site of hard conflicts over the territorial boundaries of sovereign states for a very long time (until U.S. president Trump seriously set his sights on Greenland during his second term in office), there are a range of sovereignty disputes, ranging from continental shelf delimitation issues to the interpretation of the Svalbard Treaty.

Put simply, the Arctic is undergoing fast and intensive changes. While different detailed assessments exist regarding the individual trends mentioned above, there does seems to be a consensus as to the overall developments (Dodds & Sörlin, 2022).

Of course, contrary to the somewhat polemical remark made above, nothing in this consensus should imply that further research is not necessary. Quite the contrary: while, for example, many of the central dynamics are well understood in terms of their being known to be taking place, a lot of research remains to be done to establish underlying causes and feedback processes. A prominent example here would be the study of the decreasing ice albedo, which does not seem to result from a single source of blackening, but from a regionally specific and diverse mix of black carbon, dust and ice algae. Another example would be permafrost carbon feedback and related efforts to determine carbon and methane levels in order to estimate different global emission scenarios. In addition, processes are better understood in summer than in winter resulting from a lack of year-round data sets. This holds especially true for certain regions of the Arctic which are not easily accessible (e.g. East Siberia or parts of the Canadian Arctic). However, even the most sophisticated research benefitting from an abundance of data and without accessibility issues does not necessarily translate easily into political decision-making.

To cut a long story short: while the Arctic remains a region that is changing rapidly in many respects (and with global repercussions) and though Arctic science goes on and there is a broad agreement that it should go on and that it should underpin informed political decision-making, we still lack sufficient understanding of how the entire knowledge complex of 'doing' and 'translating' Arctic science works in terms of its epistemic frameworks and their political implications. How are these shaped by—and possibly themselves shape—cognitive frames that are at least co-constituted by views that cannot directly be attributed to idealised 'neutral' motives of simply 'gaining knowledge'? What we are looking at, in other words, are both geopolitical worldviews that are 'behind', but also in a sense constitutive of, the ways in which the Arctic is practically framed in science funding, operative science and at science/politics interfaces.

1.2 Worldviews of Ice and the Formation of the Arctic

For centuries, explorers from outside the region have been fascinated by the Arctic with its harsh and inhospitable environment. The remote polar regions became a sort of place of longing for people in search of adventure, fulfilment and prestige. While Arctic communities and Indigenous peoples have been living on Arctic land, sea and ice for millennia, overwintering at research stations is still something that very few people from outside the Arctic do. Media reports, personal blogs or filmed documentaries accompany Arctic researchers in the field, and they show impressive pictures of (still) mostly frozen, white, endless, lonely landscapes. In contrast, during the summer, when research stations are booked to their full capacity, many regions turn into a green Arctic full of life. When the region is referred to today, inevitably, pictures and imaginaries are invoked that lead straight to the (seemingly simple) question: *What is the Arctic?*

While interest in the Arctic is high, the remoteness of the region provides space for worldviews. Understandings of what the Arctic 'is' depend on the different images people have of the region. Whether the Arctic is perceived and/or framed as ground zero for climate change, an Indigenous homeland, a region of opportunity for (sustainable) economic development, or as an international zone of conflict, has real implications for the political and public domain. 'Ideas matter'—and that means that worldviews influence (political) preferences and strategies. Worldviews are part of everyday communication; they are manifested in the ways things are described and the ways people act accordingly. Worldviews are based on perspectives, expectations, beliefs, values and mindsets concerning the world (or, in our case, the Arctic), which influence thought and action. As Peter Katzenstein (2022, 9) argues, worldviews help us to find our place in a world of uncertainty. Conscious or unconscious, they serve as root and framework of action, a common image to rely on:

> Relationally mediated by discourses and institutions, worldviews create narratives about what is possible, what is worth doing, and what needs to be done, as well as what is impossible, what is shameful, and what needs to be avoided. They thus have effects on the purposes and interests that shape policies and practices (Katzenstein, 2022, 9).

In other words, worldviews play a strong role in the social and spatial construction of the Arctic. On the one hand, at the individual level, worldviews can be regarded as cognitive viewpoints, shaping subjective ideas, emotions and self-perception in people's heads. On the other, and for the purposes of this study, 'they are also collective systems of thought that offer some measure of coherence and consistency in an often unfathomable world' (Katzenstein, 2022, 10). In our understanding, worldviews can encompass a range of different and more specific discursive formations such as narratives, images or frames—for example, geopolitical discourses about what is to be expected from and what has to be done about concrete power constellations; epistemic discourses defining the acceptability and the relevance of scientific knowledge; ecological discourses reflecting and shaping research priorities on permafrost, its consequences for climate change and the resulting political necessities. These discursive formations may be contested and contradictory. They may be found in a continuum, they may overlap and can possibly also be de- and reconstructed. In much the same way as Jasanoff and Kim (2015) define 'sociotechnical imaginaries', we speak of worldviews as something 'collectively held' (Jasanoff, 2015, 4). In contrast to Jasanoff, however, we think that it is an empirical question whether worldviews are 'institutionally stabilized, and publicly performed' (Jasanoff, 2015, 4). They consist of a flexible, situational, sometimes contested repertoire of expectations and values. In the daily business of making decisions and producing knowledge, worldviews provide orientation even under conditions of uncertainty. They may also fulfil a unifying function by sharing visions of (future) social orders. They mediate between single moments of action and the world. For instance, rights- and stakeholders in the Arctic follow different practices and frame the region differently. Their *collectively held worldviews* reason, legitimate and justify certain imaginaries which, in turn, orient action. Worldviews constitute the politico-epistemic basis for knowing and governing the world; they rest, however, upon a more elastic 'apparatus' (Katzenstein, 2022, 9) of condensed discursive forms. In an *institutionally stabilised* form, they can be found reinforcing themselves, but also changing, intergovernmental fora such as the Arctic Council, as well as in economic/business groups, NGOs, Indigenous Peoples Organisations, research institutions, funding agencies, international projects or shared infrastructure projects—all collaborations where a common goal, vision and understanding of the Arctic are required. *Performing worldviews publicly* aims at convincing a broader audience by evoking a set of interlinked imaginaries. This includes the ways in which strategy papers and agendas are set up, or the way that projects or results are communicated to the public.

Given the central role of science in the Arctic, we are interested in how (underlying) worldviews and related discursive formations shape the process of knowledge production—from funding, via the actual 'doing' of science, to knowledge translation at the science/politics interface. Framings of the Arctic are reproduced in a broad and ongoing practice of scientific engagement and its relevant, highly varied links to policies. Here, worldviews perform two main functions. They serve, first, as points of reference, by justifying and legitimating work, policy and practice (Katzenstein, 2022, 9) and by bringing legitimacy to funding. Second, they form a bridge enabling science and politics to work together, while at the same time being co-produced by science and politics.

The Arctic scientific community has, for instance, been very successful in creating an image of a rapidly *changing Arctic* (Steinberg et al., 2018, 166). This has various implications. On the one hand, the necessary ground for conducting Arctic research is often provided by nationally framed rationales and the resultant funding programmes. On the other, research agendas and priorities are influenced by how the natural sciences frame the objects of their study. The urgency of climate change gives scientists new authority. Moreover, it impacts the practical structures of inevitable international scientific cooperation. The translation of scientific results into policy recommendations and their implementation, however, comes with its own complexities.

It is important to note that different worldviews can exist in parallel, overlap or even stand in contrast to each other. Identifying them often requires comparing them (Jasanoff, 2015, 24–25). Looking at the extant literature on, and various (for example, visual, artistic, or media) representations of, the Arctic in general, it becomes immediately clear that the worldviews that underlie Arctic science will also come in quite different shapes and sizes. While giving a full list of such representations of the Arctic in general is far beyond the scope of this contribution, one can easily see figures like *an 'endangered Arctic', a 'global Arctic', a 'region of scientific cooperation', an 'Indigenous homeland', a 'resource frontier', a 'place for (scientific and other) adventure'*. We expect to find at least some of those, and possibly others, when we look at the worldviews in the various stages of the circle that we investigate in the chapters of this book. However, we do this by letting the empirical material speak for itself, rather than asking how much a specific worldview defined before can be found in the material. Throughout the book we are looking for references to Indigenous knowledge. We also consider the possibility that more encompassing 'Indigenous worldviews' (Hart, 2010) have an influence on Arctic science, its funding and translation into decisions. Having said that, our research is limited to exploring the science/politics interface. It cannot replace a broader study of the generation of traditional ecological knowledge, its internal differentiation and intertwinement with other forms of knowledge.

In our first case study we look for institutional, strategic worldviews which might be mirrored in funding programmes. With regard to the countries studied—Canada, Germany, Norway—we are interested in the ways in which different Arctic research funding systems are influenced by, yet also themselves partly recreate or delimit, specific worldviews. The second case study, focusing on the doing of science, turns to the individual level. Studying the Arctic scientific community in the West, and individual scientists in particular, allows us to obtain valuable insights into the framing, motivation, narratives and capabilities of the Arctic as a research object and its implications for the region. Since scientists are also people and therefore part of social groups outside the scientific world, worldviews from outside influence their working routines and therefore Arctic science itself. Socialisation and personal background, global politics, career stage and general job situation are factors which impact every individual scientist's relation to collectively held worldviews. In addition, the 'Arctic experience', that is, whether researchers have participated in expeditions or spent a considerable time living and working in the Arctic could be an important factor. In the third case study we investigate how worldviews are being translated between science and policy, using the Arctic Council as an example. In this context, we take a look at how, and to what degree, Arctic worldviews in the Council's working groups get translated into Senior Arctic Officials' reports and ministerial declarations.

1.3 Science and Politics: Exploring the Full Circle

Once Arctic science is funded and executed, the expected pathway for its results is the provision of reliable data, interpretations of the data, and recommendations for action for informed decision-making by stakeholders. However, there is a range of knowledge disputes concerning the role of science in Arctic politics, the engagement of scientists in climate issues and the relation between established science and Indigenous knowledge. From the designation of research priorities by funding agencies, to references to scientific results when particular courses of action are recommended, different worldviews permeate the interface between science and politics in the Arctic. For instance, the definition of research priorities delimits the fields and topics which will receive funding for further research. The research projects which respond to those priorities are likely to be funded, because the knowledge they deliver is considered to be of high interest by agencies at a particular moment. Once a specific topic starts to be explored, research and funding agencies tend to turn attention away from other issues that, for whatever reason, were not perceived as urgent.

At the other end, the knowledge produced by funded projects will inform the same research and funding agencies as to what should be considered future scientific priorities, in other words, which issues require further investigation. Decisions on the continuation of research priorities or on the establishment of new ones are shaped by epistemic frameworks set by the scientific research that those agencies have just funded. Likewise, knowledge produced by funded projects also aims to respond to political and societal needs defined by governments and non-governmental actors, especially if they are connected to funding and research agencies as stakeholders—policymakers envisage taking informed decisions based on the best available science. This means that, to some extent, epistemic and political discourses embedded in science funding and operation have a wider reach, shaping worldviews held by actors beyond the scientific realm. Therefore, science–politics–science forms a full circle: by advising politics, science shapes politico-epistemic worldviews; and by defining scientific priorities and funding research projects, politics both enables and delimits science, reinforcing particular worldviews and weakening others. In this sense, a seemingly 'pure' science review is already imbued with political priorities and valuations. Political worldviews are often part and parcel of the scientific evaluation process—for example, when reviewers refer to collective projects, societal problems, desirable outcomes. At the same time, political actors refer to scientific outcomes to legitimate and validate decisions.

When, in the chapters that follow, we explore this circle of Arctic science and politics, what emerges is a very complex picture beyond the linear model of science 'speaking truth to power'. The 'circle' ultimately is an ideal-typical heuristic tool that allows us to structure a reality that in fact consists of many partly independent, partly overlapping 'circles', each with its own organisational structure, dynamics, semantics, temporalities, and indeed directions. While there exists a wealth of research that deals with the science/politics interface, with the possibilities of 'science communication', with particular policy fields and with the practical and epistemological background to the scientific enterprise, we aim to analyse and shed light on a comprehensive, yet at the same time limited, power–knowledge agglomeration that is the Arctic. 'Comprehensive' because we do not limit ourselves to the individual research aspects mentioned. 'Limited' because ultimately, we can only look at a very small section of funding, research and policy.

1.4 Chapter Preview

Following a conceptual chapter, three chapters will deal with the three ideal-typical phases of the circle mentioned: 'funding', 'doing' and 'translating' science.

The conceptual chapter lays the foundations for our empirical analysis by grounding the inquiry into the circle of Arctic knowledge production and the science/politics interface in a theoretical background that we establish on the basis of various approaches primarily drawn from the fields of International Relations (IR) and Sociology. At the core we are looking for the formation and reproduction of worldviews in and between science and politics, but we also pay particular attention to important *boundary processes* and *boundary organisations* in order to explore how Arctic knowledge is validated, shared and transferred both within and between science and politics. In doing so, we need to tie these concepts back to the even broader (and older) debates on the operative autonomy of different (functionally differentiated) social systems and the resulting problems of communication. After all, not to put too fine a point on it, the frustration often heard from the side of scientists ('we have all the results, but the politicians won't listen') is not one peculiar, for example, to glaciologists working on the mass loss of the Greenland ice sheet. It points rather to basic issues concerning the autonomy of social systems as well as the possibilities and limits of directed steering. We will draw on these extensive debates in order to inquire how, and to what degree, the 'boundary work' between science and politics in our circle could be improved. We deliberately take such a conceptually rather 'deep' approach in order to forestall any assumptions that the problem at issue might be reduced to one of simply 'communicating science better' (and, most notably, hiring more communications staff or training scientists to become better communicators).

The following three chapters will each focus on one part of the circle. Chapter 3 deals with research funding, Chap. 4 with 'doing science', and Chap. 5 with 'translating science'.

Chapter 3 looks at the trends in Arctic research funding in Canada, Germany, and Norway, paying particular attention to the institutional arrangements around funding processes and the quantitative and qualitative changes in associated research outputs. A particular focus here is on the way in which interactions/dialogues between funding bodies and political actors are organised. Drawing on key strategic documents from major stakeholders in this process, we identify how four related worldviews (the *endangered Arctic*, the *Arctic as a resource frontier*, the *national Arctic* and the *global Arctic*) permeate this part of the circle.

Chapter 4 then turns to 'doing science', that is to those actually conducting Arctic research. While focusing on Germany, Norway and Canada, for practical reasons we have concentrated on the fields of permafrost and glaciological research. We are interested in the extent to which specific worldviews are present in scientific communities working in or on the Arctic. Unscrambling the *doing of science* therefore means focusing on how individual researchers conduct, reason and reflect on their research, the basis of our analysis being interviews with glaciologists and permafrost researchers at all career levels working in Canadian, German and Norwegian institutions (with 5–7 interviews for each research field in each country). The interviews seek to paint a broad picture of (world)views of and on the Arctic, asking, for example, how the Arctic is understood and framed as an object of research, how evolving Arctic research priorities are seen, how scientists themselves consider their relation to politics, and so forth. The chapter presents a multi-faceted picture of natural science in the Arctic—in particular of the scientists themselves—while also identifying a range of common threads in scientists' self-understandings as well as in the relations between politics and science.

Chapter 5 turns to the 'translation' of science to politics. In contrast to the previous chapters, it has the advantage of being able to build on a wealth of already existing research, and can thus deal with the question of how this translation takes place using a broader, yet specified approach. We look on the Arctic Council as the central nodal point in which Arctic scientific and political imaginaries combine and reinforce or prevail over one another. An important translating process takes place here when results from the working groups are discussed and taken up by Senior Arctic Officials and ministers in ministerial meetings. Although deliverables and scientific working plans are agreed, no formally binding decisions are made. Consequently, there is no clear overruling of scientific worldviews by political ones or vice versa. In this chapter, we first briefly contextualise the creation and operation of the Arctic Council from 1990 to 2020. We then focus on working groups in which glaciology and permafrost research projects have been developed, identifying through their reports which Arctic imaginaries and, consequently, which worldviews have underpinned them. We then analyse Senior Arctic Officials' reports and ministerial declarations and ask how they reflect the worldviews of scientific actors that we have identified and how they stand in relation to those of Arctic political actors. The question here is whether, and to what degree, the boundaries between science and politics are strengthened or weakened through the interplay of different Arctic imaginaries in this process.

A final chapter will bundle together our insights, reflecting both on what we have learned and on the limits of what we were able to do. In particular, we also reflect here on where, and in connection with which discourses on the Arctic, our research might stimulate further research and discussion.

Chapter 2
Knowing and Ordering the Arctic

Abstract This chapter lays the conceptual foundations for the empirical analysis by grounding the inquiry into the circle of Arctic knowledge production and the science/politics interface in a theoretical background that primarily draws on the fields of International Relations (IR) and Sociology. The focus is on the formation and re-production of worldviews in and between science and politics, but also on *boundary processes* and *boundary organisations* that play a role in how Arctic knowledge is validated, shared and transferred both within and between science and politics.

Keywords Worldviews · Science and technology studies · Research funding · Epistemic authority

In this chapter, we lay out the conceptual foundations of what we take to be the circular relationship between Arctic science and politics. The relationship is circular and not linear because the two sides influence each other by engaging in 'fictional expectations' (Beckert, 2013; Esposito, 2007) about plausible futures of the Arctic. We call these fictional but nevertheless plausible expectations, that may bridge the boundaries between science and politics, 'worldviews' (Katzenstein, 2022). After introducing the circular view and discussing the relevant literature, we elaborate more specifically on how worldviews as plausible fictions connect science and politics on a discursive level. As plausible fictions they provide actors with images, narratives or frames that envisage a future state of the Arctic, motivate their actions and work as collective points of reference for the justification and legitimation of organised ways of interlinking science and politics. While worldviews may be 'mere' fictions, they can be influential by collectively structuring the communication and practices in both Arctic research and politics. We think that this framework is crucial for an understanding of how uncertainties about the state of the Arctic are overcome, how researchers and policymakers reciprocally influence each other and find ways of coordinating their efforts even when confronted with high risks, and why various collectives of actors in science and politics seem to cling to different, sometimes even competing views and values about the Arctic. The framework will

M. Albert et al., *The Worldviews of Ice*, Springer Polar Sciences,
https://doi.org/10.1007/978-3-032-25945-5_2

provide us with a conceptual guideline for exploring the circular relationship between science and politics more closely in the subsequent chapters.

2.1 Beyond the Linear Model: Circular Relations Between Science and Politics

The relationship between politics and science has been the object of many studies in public policy research, political sociology and science and technology studies (STS) for decades, as well as, probably a bit more recently, in critical geography, environmental science and international relations (Kass et al., 2022). Proponents of a 'linear model' (Jasanoff, 2012) expect science to produce knowledge that is ideologically untainted, free from any political subtext and a more or less unquestioned basis for decision-making. Methodologically rigid evidence is supposed to be independent of interests or political values and unmistakably clear and informative for those willing to take notice of it. From this linear perspective, the role of science is straightforward: to enter the policy process at every stage via advice and expertise and to ensure that—across the different stages of problem definition, agenda setting and policy formulation, implementation and evaluation—political actors make their choices based on relevant information. In other words, the linear model rests on the assumption that the core challenge in the relationship between science and politics is for science to provide the best available evidence and to courageously present it to powerful political actors, even if it is not in line with dominating ideologies. Science is supposed to 'speak truth to power' (Wildavsky, 2007 [1979]; Hoppe, 1999).

A growing body of literature shows, however, that the figure of a more or less linear relation between science and politics cannot adequately grasp complex governance constellations characterised by a situation of high uncertainty, conflictive actor constellations and complex interactions across temporal as well as local–global scales (Scoones & Stirling, 2020). The Arctic is a prime example of such constellations. Critical geopolitics in particular has reconstructed the ways in which Arctic science takes not only an informative but also a 'formative' role (Knecht & Albert, 2022, 27). The Arctic is not simply 'a product of nature' (Ó Tuathail & Agnew, 1992), but a co-product of struggles between political and epistemic authorities, between competing geopolitical worldviews and contested understandings of validity and legitimacy. The formation of the Arctic as both a geopolitical entity and a research object can hardly be understood based on the linear model (Keskitalo, 2012).

In line with more general studies of expertise and science–society relations, recent research on Arctic geopolitics highlights the procedural and structural connection between knowledge claims and claims to be entitled to make collectively binding decisions. Instead of trying to trace the transfer and utilisation of knowledge along the 'one way street' of the linear model, we suggest reconstructing the complex and often circular interlinkages between science and politics across multiple levels and points of interaction, the formative power of both political and epistemic authorities as they reciprocally refer to each other in a search for validity and legitimacy, and the many subsequent struggles to delineate 'pure' science and political reason.

To be sure, science and politics are distinct reference systems that operate on the basis of different role constellations, actors' values, motivations and power relations (Mahony, 2013). Their boundaries, however, are demarcated by shifting actor constellations, competing discursive claims, moral positions and reciprocal expectations. The identification of such boundary dynamics is a necessary step for understanding science–politics relations and their development. Grasping the discourses and practices of 'boundary work' (Gieryn, 1998; Hoppe, 2005; Jasanoff, 2012), which at times bridge, at times separate, the two systems, is key to understanding the political and epistemic relations involved and the ways in which the Arctic is co-constructed as a politico-epistemic object.

In this often fuzzy, yet sometimes clearly demarcated relationship, the definition of scientific research goals and, consequently, the decision as to which research projects should be funded, are embedded in political discussions of what is considered relevant to be researched. Likewise, policy-making will envisage that decisions are made based on the available science, since scientific evidence is one of the most firmly established and widely accepted justifications for making policy choices in society. In other words, there is a close interlinkage or translational relationship in which science is both supported and influenced by politics, while politics is both informed and influenced by science. This circular dynamic defines the science/politics interface, and ultimately it is next to impossible to determine at what point science is no longer influenced by politics and when politics is completely 'immune' to scientific worldviews or indeed whether it ever can be. Nevertheless, this circular dynamic can only take place if both systems of reference establish clear boundaries that differentiate them from each other. Latour (2004) has pointed out how science and politics have been purposely 'distanced' by our modern effort to differentiate 'social construction' from 'nature', with the latter being praised for its independent existence which can only be translated to the social world by science. The value of science would seem to rely on its independence from the social world. What would remain special about science then, if it were not an extramundane subject, but itself embedded in social values? Likewise, how could politics justify its decisions on the basis of non-political evidence if science is perceived as politicised? The legitimacy of both science and politics requires the invisibilisation of any fuzzy interface and the demonstration of clearly defined boundaries. Sometimes circularity is subtle and the result of implicit reciprocal expectations. Sometimes it is the result of discursive pressures on science which is supposed to meet 'grand challenges' or politics which is required to be 'evidence-based' (Kaldewey, 2018; Straßheim, 2024).

The necessity to demarcate clear boundaries will depend not only on the object in question, but also on the levels of societal engagement with objects such as the Arctic. Individuals share expectations about how science—especially when publicly funded—will respond to their common concerns and needs. In that case, the presumption of an apolitical science is necessary to legitimate a political decision which claims to be based more on scientific evidence and less on the interests of competing groups. On the other hand, research objects which are relatively removed from public scrutiny—due either to their high level of complexity or to their impact on society being perceived as only indirect—will present a science/politics interface less confined by demarcated boundaries. Accordingly, when analysing the science/politics interface, both boundary work and circular translational dynamics need to be closely considered.

Our research is rooted in such an understanding of critical geopolitics, STS and theories of the circular relations between science and politics. We trace the stages in the circular process connecting knowledge- and decision-making based on Arctic research funding and Arctic research itself with the political and epistemic authority constellations in the Arctic Council. Among the important influencing factors are: the composition of research communities, agendas linked to funding agencies, and the densities of networks that allow, sustain or evoke communication between relevant communities. The literature differs on which underlying mechanisms need to be considered, and on what the political and epistemic consequences might be. Our work contributes to this discussion (1) by identifying the interaction sites at which a geopolitical object such as the Arctic is co-produced in the circular relations between science and politics and (2) by tracing the processes of translating multiple observations and claims across these sites into factual statements that become a taken-for-granted element of governance efforts.

While some authors focus on role of science in these processes (Väätänen, 2022) and others on the political tensions and constellations of constructing the Arctic (Dittmer et al., 2011), we are more interested in the co-constructions of evaluative and epistemic claims at the interlinkage between politics and science. Moreover, by exploring multiple sites of science–politics interaction we hope to get a better understanding of the gradual 'solidification' (Väätänen, 2022) of the Arctic as a geopolitical and geo-epistemic landscape.

We suggest looking at three sites of circular interlinkages and interactions: (a) Arctic research funding; (b) Arctic research itself; and (c) constellations of political and epistemic authority at the Arctic Council.

2.1.1 Preparing the Ground: Arctic Research Funding

Views on the Arctic and on what the important issues are could be expressed in the objectives, rationales and priorities of funding programmes that support Arctic research. Particularly when it comes to large-scale and costly research endeavours, it is the exception rather than the rule for funding to be driven by initiatives purely from one of the two sides, either science or politics. Arctic research funding already implies multiple mechanisms of circular influence. In 'boundary organisations' (Guston, 2001) such as the Research Council of Norway, research priorities and funding approaches are renegotiated between actors involved in both academia and ministries. In more informal or only partially formalised boundary consultations (such as, for example, the German Arctic Dialogue) views of the Arctic emerge through being articulated in specific questions and priorities. Funding is the first point at which the ground is prepared for scientific and political claims. In the administrative and financial 'hinterlands' (Law, 2004) of front-line Arctic research, funding priorities and research problems are often closely connected.

2.1.2 *Freezing the Knowledge-Scape: Collectives of Fact-Making*

Research in natural science is far from a 'pure' scientific and value-neutral exercise that is then at some point confronted with the political. Rather, research agendas and objectives are influenced by how the natural sciences, explicitly or implicitly, frame the objects of their studies in terms of underlying (geo)political worldviews. The question is the degree to which these worldviews are translated into natural science research not only in the form of underlying ontologies (what is the Arctic?), but also in the ways that practical research is organised (who cooperates with whom across national boundaries) and in what the thematic focuses of Arctic research are. We assume that different epistemic collectives determine how the inflow of facts about the Arctic is frozen into a solid knowledge-scape. 'Epistemic communities' are based on a set of shared understandings and core beliefs that form an intracommunity connection. Understandings and beliefs that are accepted by the members 'provide the glue for collective action among the individuals of the community' (Haas, 2013). Knowledge claims about causal relations in nature and society that are arrived at by consensus build a connection that also influences the interlinkage of epistemic communities with policy communities and political actors. Over time, worldviews figure as a kind of collective vantage point that brings the activities of multiple actors within the community into perspective. Sometimes these communities are held together by certain politico-epistemic tools, methods, modes of visualisation or cartographies (Simons & Schniedermann, 2021). As epistemic tools and infrastructures are constituted and shaped by multiple actors, they also actively shape and 'nurture' the collective that emerges from these activities. Sometimes these material infrastructures develop a life of their own and are closely entangled with the formation of relations between heterogeneous practices by experts and instrument designers, policymakers and civil servants, consultants and civil society organisations. Epistemic communities and their infrastructures provide plausible fictions and futures for a 'social habitat' with its own division of epistemic and policy labour, with opportunity structures for engaging in potentially new and promising fields and its very own ecology of networking and knowledge brokering across multiple levels of governing.

2.1.3 *Unfreezing/Refreezing: Struggling for Epistemic and Political Authority*

The Arctic Council, established in 1996, is the leading forum for questions pertaining to the governance of the Arctic, particularly with regard to sustainable development and environmental protection in the region. The Council is an intergovernmental forum composed of the eight Arctic states (Canada, Denmark/Greenland, Finland, Iceland, Norway, Russia, Sweden and the USA) in collaboration with a diverse network of non-Arctic states and non-governmental and intergovernmental organisations. The main work of the Council is done by its six working groups which collect and analyse data and information from national science programmes and translate them into assessment reports and policy recommendations for consideration by national governments. The Arctic Council provides multiple 'passage points' in the construction of the Arctic knowledge-scape. Experts are always moving across precarious boundaries (Blum & Straßheim, 2025). To be accepted as experts, they need to successfully claim both the competence to produce and validate knowledge ('epistemic authority') and the competence to make this knowledge relevant and justifiable for collectively ordering and evaluating society ('political authority'). Political and epistemic authority constellations, however, may change in the context of science/politics interfaces such as the Arctic Council. The success of Indigenous peoples' organisations shows that established expertise and evidence may be questioned in the light of alternative sources of knowledge (Steinberg et al., 2018). Over the course of the circular translation process, this may lead to an 'unfreezing' (or, depending on the position in expert networks, sometimes also to a 'refreezing') of the solidified knowledge-scape.

All in all, what seems to be a complex, unruly picture of a multiplicity of actors working at the boundaries between science and politics across multiple sites (funding, research, expertise) is far from chaotic. We argue that worldviews are a major ordering force in this complex landscape. They provide discursive foci which align expectations across boundaries and scales and direct them towards a future that as yet exists only in anticipation. And, even if fictitious, they motivate actors to move ahead despite all uncertainty and to collectively engage in producing the Arctic—an epistemic and political object that takes shape over the course of these various activities.

2.2 Worldviews of Ice: Plausible Fictions Between Arctic Science and Politics

Carved about 130 m inside the solid rock of a mountain on the archipelago of Svalbard lies the Svalbard Global Seed Vault, protected by permafrost. Managed by the Nordic Genetic Research Center, the facility is supposed to preserve the world's crop diversity by storing duplicates of seeds in its gene banks. More than 1.2 million seed samples of agricultural and food plants are secured in the concrete construction. The building's roof and front façade have been filled by artist Dyveke Sanne with steel mirrors of various sizes, reflecting the sun and other light. According to the artist, the installation reminds us 'of our own position in a global perspective and the conditions of our planet.' The mirrors are 'copies for a diverse landscape that demands cyclical repetition of actions, rather than continued faith in the designated original and linear progression' (Sanne, 2008). Still inviting seed donations from all over the world, the vault has become a global icon of the sciences and politics of preservation against catastrophic changes of the Arctic under the conditions of the Anthropocene (Wolff, 2021).

We call such discursive formations which are interlinking science, politics and society, based on collective anticipations of what the Arctic is and what it might become, 'worldviews' (Beckert, 2013, 2016; Katzenstein, 2022; Jasanoff, 2015; Freistein & Gadinger, 2022). By highlighting geopolitical problems, proposing possible solutions, and mobilising both evaluative and epistemological claims they shape both expectations and actions. At their core, worldviews are collective expectations about an imagined albeit plausible future. They help to distinguish between what is politically and scientifically relevant and what can be ignored at least for now, motivate actors to overcome coordination problems or societal boundaries and function as vantage points in an uncertain future that, via these anticipations, 'shape present decisions' (Beckert, 2013, 221). As Schütz and others have shown, expectations and anticipations form the core of every action (Dewey, 1929; Schütz, 1959; Koselleck, 2004). When Schütz argues, with Dewey, that action is based on an 'anticipation of future conduct by way of fantasying' he clarifies that this 'fantasying' is embedded in the knowledge currently at hand (Schütz, 1959, 84). It is an 'imagination' (Dewey, 1929) of the future made plausible by expectations formed in the present and experiences made in the past. In this sense, worldviews articulate both political and scientific aspirations. At the crossroads between Arctic geopolitics and Arctic research, however, these formative visions solidify into robust facts and normative principles that cannot easily be ignored.

2.2.1 Worldviews Between Science and Politics

In our understanding, worldviews operate on a higher level than 'mental representations' (Beckert, 2013, 220). We conceptualise them as symbolically generalised constellations of expectations (and expectations of expectations) at a collective level. They structure communication and define situations by providing interpretive guidelines for dealing with the contingencies and uncertainties of the natural and social world in a specific way (Katzenstein, 2022, 10). As plausible fictions they 'do not have to be true but must be convincing' (Beckert, 2013, 222). This means that the influence of worldviews and the orientation they offer depends on the ability of actors to maintain their validity even under conditions of high uncertainty 'as if' they were true. This is not an irrational process. On the contrary, it often involves complex calculations and theories which, in cases of disappointment, need to be corroborated by bridging explanations. In *Prisoners of Reason* S. M. Amadae

(2015) shows that game theory played a fundamental role in developing mutual expectations about the 'other side'. Maintaining non-cooperative bipolarity as an essential idea of international strategising was the result of highly scientised worldviews about rational behaviour in the Cold War.

In this sense, worldviews can be weapons (Blyth, 2002). Actors sometimes support the circulation of shared expectations that are first and foremost in their own interest. Such a 'mobilization of bias' (Schattschneider, 1960) may serve partial interests. Actors use worldviews as resources to build coalitions across multiple parties. And they even circulate them to 'attack and delegitimate existing institutions' (Blyth, 2002, 35). In his seminal analysis of macroeconomic policy-making in Britain from 1970 to 1989, Peter Hall (1993) argued that the result of changes in worldviews can be paradigmatic. In this case, the Keynesian worldview of economic policy-making was radically turned around by monetarist economic views. This shift was only possible by simultaneous changes in the epistemic authority of experts, the locus of political authorities and the overall goals of public policy. But worldviews may also stabilise authority structures and impede political as well as epistemic changes. For example, it has been argued that worldviews of environmental science dominated by Western methodologies and institutionalised within the Arctic Council's 'top-down structure' have become major obstacles to including Indigenous knowledge on an equal basis (Coote, 2023). As Jens Beckert observes, worldviews 'can prevail despite known flaws and incoherencies due to powerful particularistic interests, organizational inertia, and group pressures' (Beckert, 2013, 229).

This strategic dimension should not, however, give the impression that worldviews are merely instrumental devices in the hands of powerful interests. Such an understanding would risk underestimating how worldviews shape the direction of expectations and aspirations, relate specific ways of knowing the world to certain modes of ordering it, and provide orientation and justifications even under conditions of radical uncertainty and unpredictability. Worldviews have a performative role. Through shaping the anticipation of a predicted future, they can bring about the very actions and decisions that make this future possible in the first place. Especially in the context of economics, 'self-fulfilling prophecies' are very well known. Most prominently, the study by MacKenzie and Millo (2003) has shown that, through econometric models of financial derivatives, experts managed to change the expectations and ultimately the price development of the markets in the direction they predicted. Worldviews that manifest themselves in probability calculations and simulations may provide actors with mathematical certainty even when the future is open. Elena Esposito and others (Esposito, 2017; Kay & King, 2020) have made clear, however, that this impression of predictability is likely to be the source of sudden surprises because it can never eliminate the many 'unknown unknowns' it carries. Such moments of surprise and disappointment may also mark the beginning of revised or even brand new fictions and imaginations. Some worldviews, however, gain general acceptance and solidify in the form of institutions or material infrastructures. The Svalbard Global Seed Vault may be an example of both an institutional and material sedimentation of a worldview that is less about certainty than about reversibility in the face of a potential loss of biodiversity.

In at least three respects, the *worldviews of ice* we are concerned with in this book are specific manifestations of plausible fictions and different from other economic, political or religious worldviews. First, the worldviews at the centre of our analysis emerge under the conditions of a multiplicity of geopolitical and epistemic sources of uncertainty. In many senses, the Arctic is a fluid object that cannot be based on some kind of 'natural' geographical or geomorphological features, territorial boundaries, geological or glaciological, climatic, aquatic, or ecological dynamics alone. Critical geopolitics has argued convincingly that the determination of these 'natural' properties is closely correlated with shifts in power constellations, dynamics of social as well as epistemological inclusion and exclusion, legal claims, and cultural identities (Knecht & Albert, 2022). The degradation of permafrost has become iconic of the multiple ecological, geopolitical, and cultural uncertainties involved.

Second, these fundamental uncertainties have given rise to an unusual multiplicity of different, often competing, and contested worldviews (Dodds et al., 2013). For a long time, the Arctic has been mainly perceived from a geostrategic standpoint which manifested itself in large security infrastructures such as the former Distant Early Warning (DEW) Line, established in the late 1950s and transformed into the US North Warning System (NWS) during the 1980s. This worldview of *Arctic securitisation* persists in climate modelling efforts carried out by the US military in order to estimate the damage to military infrastructures of the NWS and detect any danger of an 'Arctic-originating attack'. This worldview of Arctic securitisation stands in stark contrast to a focus on the *Arctic as Indigenous homeland,* its post-colonial structures, the Indigenous peoples' territory, and their local self-determination. Torn between starkly contrasting worldviews such as these, the Arctic is permanently re- and co-constructed, and these processes of Arctic co-construction have manifest consequences in local as well as international constellations. Tracing the worldviews of ice is also essential for understanding how the Arctic is being shaped and what the consequences are.

Third, worldviews of ice emerge and solidify at the interlinkages between science and politics. Depending on the political and epistemological values incorporated in it, any worldview has implications for the boundaries between science and politics. For example, 'speaking truth to power' as an influential politico-epistemic narrative is rooted in a worldview based on a clear separation between the two spheres and the expectation that evidence-based policy results in better decisions. This conviction is also the foundation of Arctic policy guidelines across multiple countries (Gehrke & Wehrmann, 2024). In other institutional contexts, such as the Arctic Council, a worldview based on co-production seems to have become more influential. The Arctic Council's Indigenous Peoples' Secretariat (IPS) not only represents the interests of Indigenous peoples, it also manifests the formal recognition and incorporation of Traditional Knowledge/Indigenous Knowledge as an essential way of understanding the Arctic and its physical, biological and cultural resources.

To reconstruct both the changes in Arctic worldviews and the influence they exert, we focus especially on the formation, manifestation and contestation of worldviews between science and politics across the three interaction points of funding, research and cooperation at the Arctic Council.

2.2.2 *Structure and the Analysis of Worldviews*

While we provide an analysis for which we use concepts that are strongly informed by theories, we do not aim to intervene in theoretical debates on concepts themselves. Needless to say, if we wanted to contribute a theoretical discussion of the concept of worldviews and, most notably, 'imaginaries' (which would, by extension, also cover 'discursive formations', 'communities of practice', etc.), that would warrant a book of its own.

We use concepts as heuristic tools; most prominent among them being the concept of a 'worldview' as used here. On a general level, and following Katzenstein's (2022) proposal, we see them as socially constructed frameworks that reduce uncertainty by providing a shared sense of the possible and the impossible, as well as of the desired and undesired. We add to this the specification that, in principle, multiple different and possibly overlapping worldviews can exist at any given time. Individual worldviews can be specified by their relation to a common repertoire of referents (in the semiotic sense) that distinguish them from other significant worldviews and the associated referents in their environment. In our case this repertoire of referents is the Arctic as a region that is symbolised by its (partial) ice cover. We see the always necessary, if broad, specification of worldviews as marking a specific difference from their closely related cousins, that is, imaginaries, which serve as transmission belts between many individual and collective representations of worldviews, expressed through various forms of visual or textual expression.[1]

Nonetheless, we do not wish to dwell too long on the relation between worldviews and imaginaries. Since they are concepts defined in quite different ways, as well as in different intellectual traditions, their relation often remains unclear even in works that use both.[2] This 'conceptual ambiguity' has also caused significant methodological problems (Bollman, 2022, 1). The empirical reconstruction of worldviews or imaginaries often seems to follow highly diverse principles which are not always made explicit. Methodological complications occur when definitions are lacking, conceptual specifications are fuzzy, and the underlying criteria seem arbitrary. To deal with these methodological challenges, we follow a pragmatic approach suggested in some of the recent literature by developing our conceptualisation of worldviews and their core components by analogy with the conceptually matured, empirically fruitful tradition of research on framing (for a similar approach see Bollman, 2022).

Very much like worldviews, frames were originally defined by Erving Goffman (1974) as 'schemata of interpretation' enabling actors 'to locate, perceive, identify

[1] Here we draw primarily on Castoriadis's (1997) understanding of the 'imaginary institution' of society; see also Albert (2024). In terms of practice theory, one could here also refer to 'general understandings' as a condensed expression of widely shared imaginaries.

[2] Katzenstein remarks that worldviews 'are imaginaries that are built around basic and often unarticulated assumptions' (Katzenstein, 2022, 15), which does seem to be a definition quite difficult to specify further. On the discussion on imaginaries see also Albert (2024).

and label' events, organise individual as well as collective experiences and expectations, and guide actions within everyday life and the world at large. The literature on geopolitical social movements has borrowed this notion and developed it into a multidimensional framework of 'frame tasks' or 'frame elements' that has since been used across a wide range of empirical cases (Entman, 1993; Snow & Benford, 1988). We loosely draw on the approach originally proposed by Snow and Benford (1988) in their seminal analysis of protest mobilisation frames. This approach describes the internal structure and dimensionality of frames, proposes a fixed set of criteria and elements, and develops assumptions about how they are related to each other. Worldviews and frames are conceptualisations on different levels that come from different theoretical traditions. They must not be used interchangeably. The approach to structuring and analysing frames developed by these scholars can, however, serve here as a model for developing a heuristic that structures our understanding of worldviews (see below).

The worldviews at the focus of our analysis are more specific than frames and, at the same time, broader in terms of their socio-political horizons. They are more specific inasmuch as they do not refer to all kinds of meanings and interpretations in everyday life, only to those which are produced and reproduced at the interlinkages between science and politics. They relate epistemic and political authority to each other in a specific way, assign meaning to the boundaries between science and politics, and condition the procedures by which epistemic claims are validated and translated into politically relevant knowledge.

At the same time, the worldviews that we are interested in would usually be expected to be 'geopolitical' ones. They are situated within a social and political horizon much broader than the issue-specific frames analysed, for example, in policy research or media studies. We assume that where a worldview relates to something that necessarily has a spatial dimension—and this is undoubtedly the case with the Arctic, however its boundaries may be defined in detail—it will always contain references to a concept of 'world order'. That concept could be specified in vastly different ways, but will usually include at least some understanding of how the world is ordered in political terms, and that understanding will, in turn, usually include some understanding of a geographical figuration underlying or expressing that order. These understandings are not fixed, but often deeply sedimented.

This in a sense pretty much sums up what since the 1990s has emerged as the research tradition of 'critical geopolitics' (Dodds et al., 2013). Critical geopolitics seeks to analyse the roles that understandings of geographic figurations play in shaping political spaces, as well as to enquire into how, in this context, political spaces are embedded in a wider environment (such as, for example, in relation to underlying assumptions about nature–society relations). The basic 'critical' idea here refers to 'classical' geopolitics, which essentially assumes a law-like relation between geographical features and power-political configurations. A critical geopolitical perspective does not deny that there are deeply sedimented assumptions, rooted in a variety of worldviews, about what world order is and how it is shaped, but it asserts that, ultimately, these assumptions and worldviews are contingent on historical and social context and do not reflect quasi-natural laws.

Informed by frame analysis we propose a threefold structure for worldviews. It will help us to identify them empirically, unravel their internal structure, and get a better understanding of their coherence and inherent tensions. The proposition is that worldviews structure the interaction between science, politics and society by (1) diagnosing geopolitical problems and their causes, (2) forming expectations about adequate solutions and their implementation in the future, and (3) making evaluative and knowledge-related claims regarding the epistemological validity and geopolitical legitimacy of these diagnoses and expectations. The underlying assumption is that the way these three interpretive 'tasks' relate to each other within a worldview, as well as between potentially overlapping (or competing) worldviews, motivates and constrains the interactions between Arctic research and politics, mobilises understandings of how to analyse and order world relations, and ultimately leads to the collective formation of the Arctic as a geopolitical and geographical object.

Table 2.1 summarises these structural elements of worldviews and illustrates them empirically using the example of the *endangered Arctic* worldview. We use the

Table 2.1 Structural elements of worldviews

Interpretive task	Worldview element	Description
Diagnosis of geopolitical problems and their causes	Topic or theme	Central issue and its relevance for research and decision-making; examples: the Arctic as 'ground zero' for climate change
	Actors	Main entities to which agency is ascribed (individual or collective); examples: global political and economic actors; transdisciplinary researchers; NGOs; Indigenous peoples; polar ecosystem species
	Causal relations	Cause–effect relations and underlying mechanisms; examples: global biodiversity and biogeochemical cycles/tipping points of polar ecosystems; tragedy of the commons; international relations dominated by interests of states and industry
Prognosis of solutions	Solutions	Courses of action, related strategies, tactics, targets and modes of evaluation; examples: global ocean treaty; environmental controls; sustainable research and science communication; public protest; emotional identification with the Arctic
Values and epistemological foundations	Legitimacy of action and core values	Geopolitical legitimacy of action; examples: reaching tipping points of world climate; saving global ecosystems 'before it is too late'
	Attribution of benefits	Who benefits? examples: ecosystem stability, Arctic peoples, world climate
	Risks and opponents	Constructions of friends/foes; attribution of risks; examples: state–industry complex, irresponsible consumption, military interests
	Science and politics	Modes of constructing science–politics relations and underlying values; examples: deliberative model of science advice; co-production; epistemic diversity; transdisciplinarity; evidence-based policy

table as a template for the identification and analysis of worldviews based on a qualitative, category-based content analysis of documents such as funding guidelines, research reports or policy briefs (Matthes & Kohring, 2008; Bollman, 2022, 4).

As the next chapters show, interpretive tasks are not necessarily related to each other in a coherent way; diagnostic elements may be strong, but solutions may be underdetermined; worldviews may overlap or, by contrast, emerge in strong competition to each other. The composition, internal structuration and external relations of worldviews are important features of analysis. They may affect their geopolitical relevance, epistemic credibility and potential to mobilise action by those who identify with them—and those who reject them (Snow & Benford, 1988, 205–211).

Chapter 3
Constructing the Arctic I: 'Preparing Science'

Abstract This chapter looks at the trends of Arctic research funding in Canada, Germany, and Norway, paying particular attention to the institutional arrangements around funding processes as well as on quantitative and qualitative changes in associated research outputs. A particular focus is on the way in which interactions/dialogues between funding bodies and political actors are organised. Drawing on key strategic documents of major stakeholders in this process, four related worldviews are identified ('endangered Arctic', 'Arctic as a resource frontier', 'national Arctic', 'global Arctic') that permeate this part of the circle.

Keywords Research Funding · Funding Organizations · Publications · Canada · Norway · Germany

3.1 Preparing Arctic Science: Structures and Trends in Research Funding

This chapter reconstructs the development of Arctic science in three national funding systems and explores how worldviews are mirrored in the designation of research priorities by funding organisations. Based on an analysis of funding and publication data on the one hand and key strategic documents issued by central funding organisations on the other, we provide an overview of the development and institutional embeddedness of Arctic science funding in Canada, Norway and Germany since the early 1990s. We outline similarities and differences between the funding systems and the role that funding organisations play in different science–politics consultation formats.

Qualitative studies on the significance of research funding in Arctic science and policy-making are rare. This contrasts especially with the wealth of studies devoted to science diplomacy that examine Arctic science more generally as a 'soft power' fostering national foreign policy interests (Ruffini, 2020; Wood-Donnelly & Bartels, 2022, 2). However, considering that 'Arctic and non-Arctic states have historically

M. Albert et al., *The Worldviews of Ice*, Springer Polar Sciences,
https://doi.org/10.1007/978-3-032-25945-5_3

used, and continue to use, scientific activities for foreign and security policy reasons' (Bertelsen, 2020, 242), Arctic science is by no means a process within a 'neutral infrastructure' (Skupien & Rüffin, 2020, 25). It is influenced by rationales of various geopolitical worldviews at the boundaries of science and policy-making. Arctic science funding is therefore not just a technical issue in the scientific process of knowledge production, it is also a social 'ordering process' (Jasanoff, 2005), and an important element when it comes to understanding the boundaries between knowledge production and policy-making.

The literature in science studies and research policy provides a variety of approaches to examining the role of research funding at the boundary of science and policy-making and emphasises different aspects. Some scholars approach the science–politics boundary as an institutionalised structure, for example by treating funding agencies as mediating and stabilising boundary organisations between science and policy (Braun, 1993; Guston, 1999; Braun & Guston, 2003; Klerkx & Leeuwis, 2008; Gustafsson & Lidskog, 2018). Also, the dynamics of social processes and various actor constellations are described as different forms of coordination modes and social ordering processes within certain institutional framework conditions (van der Meulen, 1998; Lepori, 2011; Jasanoff, 1987, 2005). Others emphasise the dynamic social processes of 'boundary work' (Gieryn, 1983) where authority in science–politics relations as well as the balance between knowledge production and policy-making are constantly renegotiated (Guston, 2000; Hoppe, 2005; Jasanoff, 2012; Jungcurt, 2013). Especially from the perspective of the relevant actors in science themselves, the boundaries to policy-making are 'drawn and redrawn in flexible, historically changing and sometimes ambiguous ways' (Gieryn, 1983, 781).

Whether boundaries are conceptually described as institutional framework conditions or social processes, the various rationales of the Arctic as a political, cultural and economic region lead to varying understandings 'of the point at which the autonomy of science ends and the role of political decision-making begins' (Jasanoff, 1987, 200). Any analysis of Arctic 'knowledge regimes' (Campbell & Pedersen, 2008) needs to look beyond the political and economic framework conditions of nation states and also take into account specific cultural and historical perceptions of the Arctic as a region. Earlier studies on Arctic science funding indicate that Arctic and non-Arctic states differ significantly in their practices with regard to reporting Arctic-related scientific activities, outlining scientific programmes, and establishing funding procedures (Łuszczuk et al., 2020, 618), something which, in our view, cannot be sufficiently explained by different institutional arrangements alone. A variety of domestic and foreign policy 'drivers' and objectives influence the understanding of Arctic science in different countries (Andreeva & Hønneland, 2023). Thus, studies of Arctic science funding need to consider both institutional framework conditions as well as specific social and cultural developments and perceptions at science/politics interfaces. Worldviews are therefore important for a better understanding of the characteristics of Arctic science funding.

When outlining the specific characteristics of Arctic science and its funding in Canada, Norway and Germany, we limit the focus to central public funding organisations which (1) provide accessible data on funding input to Arctic science since the

1990s and have a publication output that can be linked to them, and (2) play a significant role in established consultation formats at the science/politics interface. As a result, the central funding organisations considered in this chapter are the Research Council of Norway (RCN), the German Federal Ministry of Education and Research (BMBF), the German Research Foundation (DFG), the Natural Science and Engineering Research Council (NSERC), and the Social Science and Humanities Research Council (SSHRC) in Canada. While these are not the only significant funding organisations in Arctic science in the three countries examined, their organisational structures and established funding practices represent key similarities and differences in Arctic science funding across various national research funding systems.

The chapter starts by reconstructing the development of Arctic science since the 1990s. We outline the established practices of public funding organisations as central actors in preparing Arctic science and show general funding trends. Canada, Norway and Germany share a long tradition in Arctic science, characterised by established organisational structures as well as research and funding practices. They also share some general perceptions of the significance and function of Arctic science, particularly with regard to climate change. However, they differ in terms of how the Arctic is perceived as a geopolitical, cultural and scientific region. As the title of the chapter indicates, preparing Arctic science also includes the construction of the Arctic as a region of scientific and political significance to justify publicly funded Arctic science. Worldviews fulfil an important function by providing orientation and points of reference in this respect. A central aim of this chapter is therefore to analyse which worldviews appear in key documents on Arctic science in the three countries, and to identify similarities and differences between these worldviews. In the following sections, the central funding organisations and the development of Arctic science funding and science-politics consultation formats from 1990 to 2020 are compared (Sect. 3.2). A short methodological section explains our approach to analysing the key documents of funding organisations (Sect. 3.3). The main findings from the analysis are then discussed with regard to the structural elements of worldviews as laid out in Chap. 2. These include considerations of geopolitical problems, prospects for solutions, and epistemological foundations (Sect. 3.4). The chapter concludes with a short summary and a table of the main worldviews operative in preparing Arctic science (Sect. 3.5).

3.2 Funding Arctic Science in Canada, Norway and Germany Since the 1990s

3.2.1 Main Funding Organisations

Governmental organisations play a major role in Arctic science funding. In Canada, the relevant agencies include two federal ministries: the Department of Foreign Affairs and International Trade (DFAIT) and the Ministry of Crown–Indigenous Relations and Northern Affairs (CIRNAC), as well as the three territorial governments of Nunavut, Northwest Territories and Yukon. In Norway, all ministries are involved in Arctic science in their respective sectors, most significantly the Ministry of the Climate and Environment (KLD) and the Ministry of Education and Research (KD). In Germany, the most important federal government ministries involved in Arctic science funding are the Federal Ministry for Education and Research (BMBF)[1] and the Federal Ministry for Economic Affairs and Energy (BMWE).

The role of ministries and other governmental agencies operating as funding organisations varies significantly across the three countries. In Canada's research funding system, the most important organisations are the three research councils: the Natural Sciences and Engineering Research Council of Canada (NSERC), the Social Sciences and Humanities Research Council of Canada (SSHRC), and the Canadian Institutes of Health Research (CIHR). While the first two were established in the 1970s, the CIHR was founded only in 2000. All three councils operate as federal funding agencies to support post-secondary research in Canada, including Arctic science. By contrast, the Research Council of Norway (RCN) is the only national research council in this country. It was established in 1993 after five research councils were merged into one (Skoie, 2000). The RCN manages a large part of public research funding by the government, supporting basic and applied research across sectors and disciplines (RCN, 2019; Kallerud et al., 2001). Germany, by contrast, has no federal research councils. Instead, research associations like the Helmholtz Association (HGF) consist of numerous non-university research centres, most notable for Arctic science is the Alfred-Wegener-Institute Helmholtz Centre for Polar and Marine Research (AWI). Other Helmholtz centres, such as GEOMAR Helmholtz Centre for Ocean Research Kiel and the GFZ Helmholtz Centre for Geosciences, also significantly contribute to Arctic science. In total, the HGF consists of 18 research centres which 'pool their expertise in long-term, cross-centre programs' (Helmholtz, 2026). The so-called Program-Oriented Funding (PoF) is described as a 'strategic framework' for the research association to set 'the priorities for Helmholtz's research and its funding over program period' (Helmholtz, 2026). In this programme funding mechanism within the HGF, funds are allocated in a competitive process based on both 'scientific and strategic criteria', evaluated by international review panels (Helmholtz, 2026). For the AWI, as the central coordinating research centre for German polar research, 90% of its core funding comes from PoF. The remaining 10% of the core funding comes from the federal states in which the AWI has facilities (Bremen with 8% and Brandenburg and Schleswig-Holstein with 1% each) (AWI, 2020). In addition to the more strategic oriented programme-funding, the BMBF also provides individual project grants to the AWI and other research centres and universities. Most of these project funds from the government are channelled through the Projektträger Jülich (PtJ), which operates as a limited company and non-profit organisation managing both public research funding and evaluation processes for the federal government, including the project funds provided by the BMBF and the European Commission. A second important pillar in

[1] Following a government reorganisation in 2025, most parts of the ministry were integrated into the new Ministry for Research, Technology and Space (BMFTR). For coherence we continue to refer to the BMBF.

German polar research funding is the German Research Foundation (DFG). As a non-profit organisation, it provides project funding particularly for university-based research.

Overall, the main funding organisations in Arctic science differ in terms of their institutional formats as governmental or non-governmental organisations, as well as in their level of centralisation. In Canada and Norway, research funding is centralised in governments and public agencies, while in Germany governmental and non-governmental organisations operate separately as autonomous funding entities.

3.2.2 Funding Trends from 1990 to 2020

3.2.2.1 Data Collection Process and Limitations

A central challenge when tracing Arctic science funding arises from the fact that reporting styles are mostly not standardised, and that they vary not only between countries, but also between individual organisations. That said, Norway and Canada have established quite transparent reporting systems for their research funding. For example, the funding organisations operate publicly accessible online project databases, such as the NSERC's Awards Database,[2] the SSHRC Awards Search Engine,[3] and the RCN Project Databank.[4] German funding data is less accessible. The DFG database 'GEPRIS'[5] provides general information on the number of funded projects, but not on the amounts of funding they receive. Project funds from the BMBF are published in annual reports of the PtJ. Another major challenge to investigating Arctic science funding, which has also been noted by Dag Aksnes and others (Aksnes et al., 2023; Osipov et al., 2017), lies in the inconsistency in the definition and operationalisation of 'Arctic science'. Administrative regulations for allocating and evaluating research funding, as well as decisions regarding what is funded as Arctic science, will most likely continue to be made at national or provincial level, meaning that international comparative studies on Arctic science funding will remain problematic. The UArctic Thematic Network on Arctic Research Analytics and Bibliometrics, hosted by Umeå University,[6] provides the most comprehensive reporting on Arctic science and funding trends. The aim of the network is to identify

---

[2] NSERC *Award Database*. https://www.nserc-crsng.gc.ca/ase-oro/index_eng.asp [accessed 23 Jul 2025].

[3] SSHRC *Awards Search Engine*. http://www.outil.ost.uqam.ca/CRSH/RechProj.aspx?vLangue=Anglais [accessed 23 Jul 2025].

[4] Research Council of Norway *Prosjektbanken*. https://prosjektbanken.forskningsradet.no/en [accessed 31 Aug 2025].

[5] Deutsche Forschungsgemeinschaft (DFG). *GEPRIS—Projects funded by the DFG*. https://gepris.dfg.de/gepris/OCTOPUS?language=en [accessed 09 Sep 2025].

[6] Established in 2015 as the UArctic Science & Research Analytics Task Force (Aksnes et al., 2016, 3).

knowledge gaps and the main challenges for Arctic science through big data analysis using bibliometric and scientometric methods. It has developed sophisticated keyword searches for geographical and Indigenous peoples' names and terms, in addition to general terms like 'arctic', 'polar' and 'north'. Furthermore, the network employs a broad definition of 'Arctic science', referring primarily to the definitions as suggested by the Arctic Council's Working Groups (Aksnes et al., 2024, 8).

Despite these methodological challenges, we collected a dataset based on UArctic bibliometric trend analyses, project databases provided by funding organisations, as well as the Scopus abstract and citation databases provided by the academic publisher Elsevier. Analysis of the collected dataset reveals important similarities and differences in Arctic science funding in terms of both the inputs of funding and the output of publications linked to Arctic science. Canada, Norway and Germany show different approaches to Arctic science regarding dedicated budgets and the number of projects funded. Previous bibliometric studies have indicated that between 2016 and 2022, Canada was among the top three countries worldwide in terms of the number of funded research projects associated with Arctic science, while Norway spent the highest amount after the US and the European Commission. In the same report, Germany ranks among the top ten countries in terms of projects funded, although the funding amount is relatively low in comparison (Aksnes et al., 2024, 18).

Using the data collected from funding organisations, we explored whether, despite differing reporting styles and currencies, it was possible to identify similarities and differences in Arctic science funding trends. To this end, we searched for projects funded between 1990 and 2020. We did not develop overly detailed search queries for 'Arctic science', because the term has varying and inconsistent meanings. Seemingly precise search queries could lead to statistical artefacts. Instead, we opted for a relatively broad search strategy and only added key words such as 'north' and 'northern' where relevant, as in the cases of Canada and Norway, to capture perceptions of the Arctic as a global or local region respectively (see Sect. 3.4). Where possible, the number of projects and funding amounts were determined according to the year in which they started. Data sets based on fiscal years list multi-year projects for each year separately, which distorts the actual funding amounts and number of projects for the year in question. In addition, due to discrepancies in the presentation of the figures, the minimum funding amount for the NSERC and SSHRC was set at CAD 10,000. This means that some relevant data is no longer included in the analysis. Comparing our results to previous trend analyses (Aksnes et al., 2016, 2023, 2024; Osipov et al., 2017), however, we did not recognise significant deviations in the general funding trends, making the margin of error acceptable. After collecting the data, the results were double-checked manually, and projects with no obvious reference to the northern circumpolar region were minimised as far as possible. Examples of such false hits included project topics that referred to the 'northern' parts of other countries or regions (e.g. northern African countries or Northern Ireland).

The results presented in this chapter are based on a total of 4877 hits for NSERC and SSHRC, and 1186 for the RCN in their respective project databases. The numbers for the DFG were collected from its 'GEPRIS' database, and funding data for the BMBF were collected from available PtJ annual reports between 2010–2020, as well as from previous bibliometric trend analyses (Osipov et al., 2016; Aksnes et al., 2023, 2024). Because of the variance in the quality and accessibility of the data, the tables in this section can only illustrate general trends in Arctic science funding. In the data collection process itself, the methodological and technical challenges mentioned previously, along with the search for comparable parameters for Arctic science, revealed the different funding practices and reporting approaches.

3.2.2.2 Funding Inputs

Canada, Norway, and Germany show varying traditions of embedding Arctic science either in higher education institutions (HEIs) or in research institutes outside universities. For instance, in Canada, the NSERC and SSHRC are mandated to primarily fund university-based research. Both project databases reveal that over 90% of funding goes to the HEI sector, in terms of the numbers of projects and the amount of grants. The NSERC is the most significant funder of Arctic science in Canada. Our data showed three funded projects in 1991 worth around 400,000 USD and 239 projects worth 8.7 million USD in 2020.The SSHRC funded 11 projects in 1998 (200,000 USD), which increased to 157 projects (4.7 million USD) in 2020. The share of projects funded by the third research council, CIHR, is comparatively small, and data is less accessible. However, the significance of health-related Arctic science is expected to increase. CIHR is involved in various tri-council research and funding initiatives, such as the Networks of Centres of Excellence (NCE). These network schemes, as we will see below, are central funding mechanisms and modes of coordination for Canada's research funding system and for the development of Arctic science in Canada (Lepori, 2011).

In Norway, the funds allocated by the RCN for Arctic science increased from around ten million USD in 1996 to around 30.8 million USD in 2020. The number of projects, however, remains significantly lower compared to the NSERC in Canada. The data collected from the project database of the RCN reports one project in Arctic science for 1996. The number of projects peaked in 2013, with 138 funded, and decreased slightly after that to 104 in 2020 (Fig. 3.1). One explanation for this is that a large portion of RCN project funding is allocated to research initiatives and programmes lasting for several years, such as the Centres for Research-based Innovation (SFI) or the Norwegian Centres of Excellence scheme (SFF) (RCN, 2025a, 2025b). Figure 3.2 illustrates how the RCN has a funding cycle in which outputs peak every 3–4 years, followed by years of lower funding inputs. The funding practice in the other two countries indicates less but more constant funding. Furthermore, a report from 2018 by the Nordic Institute for Studies in Innovation, Research and Education (NIFU), mapping Norwegian polar and high north research, indicates that only around 25% of Arctic science funding is allocated by the research councils' funding programmes. Around 75% of Norwegian Arctic-related research funds are allocated as direct funding by ministries for public agencies or

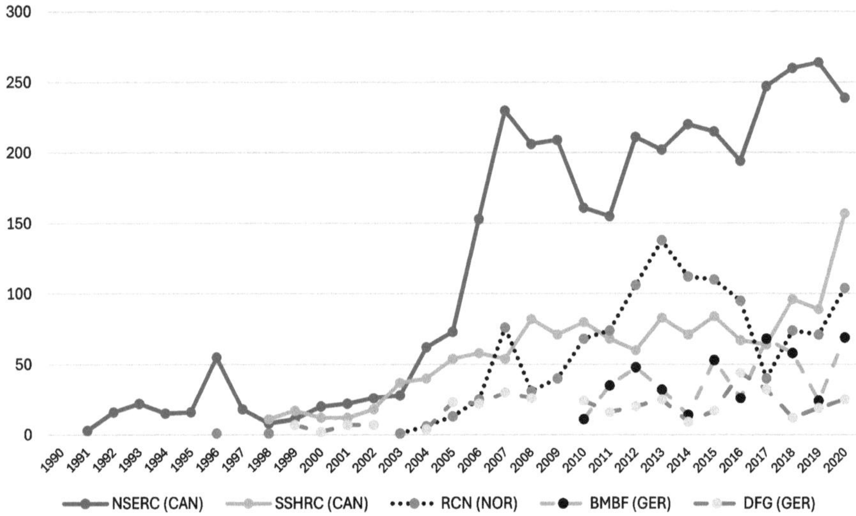

Fig. 3.1 Number of projects by starting year between 1991–2020. (Data collected from: NSERC's *Award Database* [accessed 23 Jul 2025]; SSHRC *Award Search Engine* [accessed 23 Jul 2025]; Research Council of Norway *Prosjektbanken* [accessed 23 Jul 2025]; Projektträger Jülich. *Geschäftsbericht* [accessed 23 Jul 2025].)

departments like the Norwegian Polar Institute (NPI) (Aksnes & Rørstad, 2019, 36 and 53).

In Germany, most Arctic-related research funding is provided by the BMBF and the DFG. However, Arctic science is not reported on explicitly but rather as part of German polar research, which, in turn, is part of the larger marine and polar research programme. The Alfred-Wegener-Institute (AWI) is the central embodiment of this approach (Salewski et al., 2020). Accordingly, the German funding amounts presented here are relatively high in comparison to the numbers reported by Canada and Norway. Annual reports from the PtJ reveal a slight decrease in the budget available for polar research projects funded by BMBF, from around 30 million USD in 2010 to 20.1 million USD in 2020 (Fig. 3.2). At the same time, the actual number of project starts funded by the BMBF fluctuated but, overall, increased from 11 in 2010 to 69 in 2020 (Fig. 3.1). However, the data here provide for a somewhat narrow view of Arctic-related research funded by the BMBF. There are various explanations to describe the variance in project-based research funding and the actual number of projects. One aspect which is not shown in our data is the growing importance of EU and other trans-national research funds as well as international research projects with various project partners and funding sources from different countries, where the BMBF provides only parts of the funding. Furthermore, and similar to Norway, project funds make up only a small part of Arctic science funding in Germany. For example, in the case of AWI the PoF accounts for a large part of its total research budget (90%). Since the establishment of the PoF in 2001, the budget has increased quite significantly over the past decades. Based on annual reports by

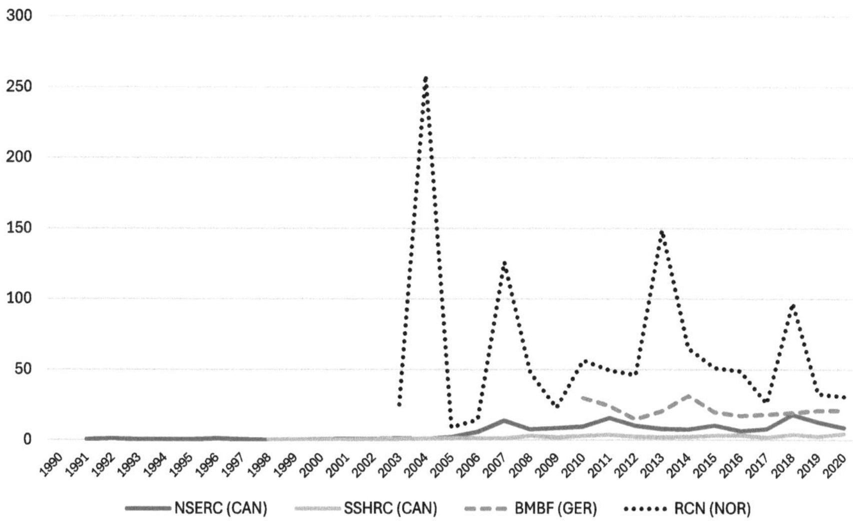

Fig. 3.2 Funding trends in Arctic science between 1990–2020 in million USD. (Data collected from: NSERC's *Award Database* [accessed 23 Jul 2025]; SSHRC *Award Search Engine* [accessed 23 Jul 2025]; Research Council of Norway *Prosjektbanken,* [accessed 23 Jul 2025]; Projektträger Jülich. *Geschäftsbericht*, [accessed 23 Jul 2025]; For a better overview and readability, currency figures (CAD, NOK, EUR) are converted into US Dollars. The exchange rate was adjusted for the relevant year.)

AWI and the Helmholtz association, since 2003 the PoF for AWI increased 2003 from around 91 million USD (79 million EUR)[7] to around 170 million USD (152 million EUR) in 2020.[8] University-based Arctic science, in turn, is largely funded by the DFG through individual research grants. The DFG project database does not report funding amounts, but Arctic-related projects can be traced back to 1999. The data shows seven projects began in 1999, increasing to 25 in 2020. The highest number of 44 funded projects was recorded in 2016. The total number of DFG-funded projects between 1999 and 2020, based on its own project database, stands at 371 (Fig. 3.1).

Overall, our data analysis supports the findings of previous bibliometric studies, showing that Canada funds a large number of shorter one-to-two-year research projects in HEIs with relatively small budgets, while Norway's Arctic science funding favours fewer but longer and larger-scale projects, with correspondingly greater budgets (Osipov et al., 2016, 13; Aksnes et al., 2024, 18). The number of funded projects and amounts indicate significant differences in the general funding approach

[7] First report of PoF budget for AWI in Helmholtz annual report 2006 covering the budget since 2003 (Helmholtz, 2006, 76).

[8] PoF budget for AWI in 2020: 152 million EUR (BMBF 90%, state of Bremen 8% and states of Brandenburg and Schleswig-Holstein 1% each) (AWI, 2020).

to Arctic science. Looking further to the output of publications related to Arctic science gives us an idea of thematic trends in the subject since the 1990s.

3.2.2.3 Publication Output

In order to identify possible changes and general trends in Canada, Norway, and Germany, we are now looking more closely at trends in the subject areas of Arctic science. A document search in Scopus with the keyword 'arctic' for the time period 1990–2020 shows that the central funding organisations document an increased publication output for Arctic science. In total we identified 5826 Arctic-science-related publications (RCN: 2342, SSHRC: 195, NSERC: 1839, BMBF: 505, DFG 945). We acknowledge that our search parameters are not as refined as those used in the large-scale bibliometric studies of the UArctic Thematic Network. Nevertheless, the results offer some insights into which subject areas are funded in Arctic science. A pilot study on Arctic Research Publication Trends from 2016 shows that Arctic-science-related publications increased globally from around 5000 in 1996 to over 11,000 in 2015 (Aksnes et al., 2016, 13). The relative proportion of Arctic science in the global scientific publication output thus remains low and stable at about 0.35 to 0.42% (Aksnes et al., 2016, 14). While the relative global output of Arctic science publications remains relatively stable, the total number of publications associated with individual funding organisations shows a notable increase since 2015 (Fig. 3.3).

Between 1990 and 2015 the increase in publication output is relatively low, which might be also due to fewer publications being recorded in the databases. After

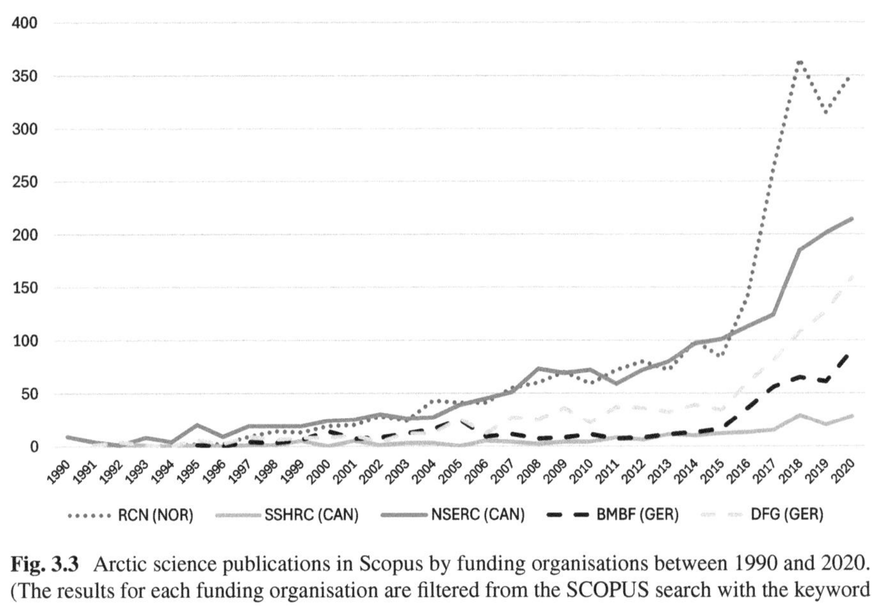

Fig. 3.3 Arctic science publications in Scopus by funding organisations between 1990 and 2020. (The results for each funding organisation are filtered from the SCOPUS search with the keyword 'arctic' and time period 1990–2020.)

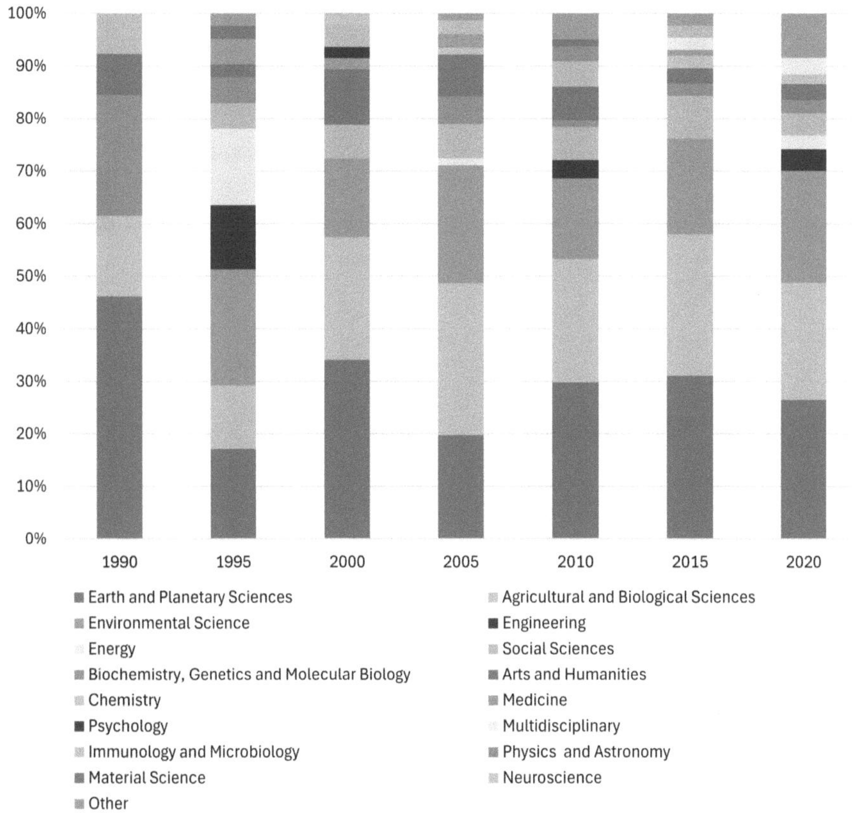

Fig 3.4 Publication subject areas funded by NSERC and SSHRC between 1990–2020. (Publications funded by the NSERC and SSHRC between 1990 and 2020. Documents by subject areas in SCOPUS.)

2015 the output increased significantly in all three countries. The highest number of publications associated with funds from the RCN was recorded in 2018 when there were 365 publications, as well as 29 from the SSHRC. For the other three funding organisations the peak year is shown to be 2020 when NSERC funded 214, the DFG 159, and the BMBF 90 publications (Fig. 3.3).

There are also noticeable similarities regarding the prioritised subject areas, indicating similar funding priorities for Arctic science. In all three countries, Arctic science is primarily focused on research in the natural sciences. Accordingly, the top three subject areas, as defined by the Scopus database, are earth and planetary sciences, environmental sciences and agricultural and biological sciences. This means that Germany, Canada and Norway have similar priority settings in Arctic science, which are closely connected to climate and environmental changes in the region. Figures 3.4, 3.5, and 3.6 show that the top three subject areas remain constant, with two exceptions in Germany and Canada in 1990. Furthermore, while the funding organisations in Norway and Germany support all subject areas, Canada

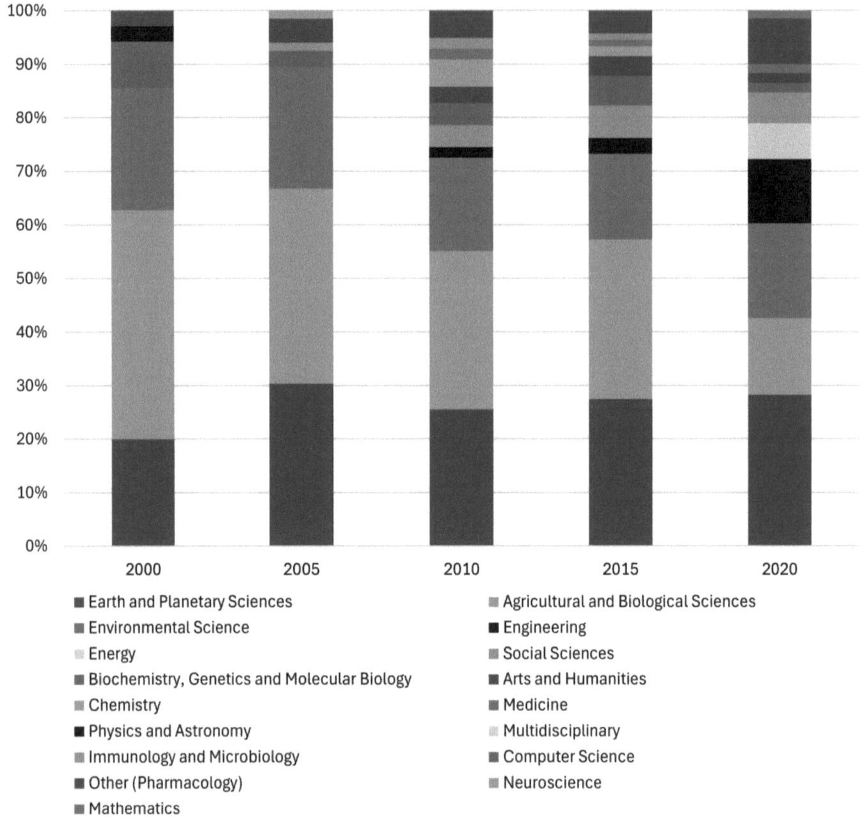

Fig. 3.5 Publication subject areas funded by the RCN between 2000–2020. (Publications funded by the RCN between 2000 and 2020. Documents by subject areas in SCOPUS. Publications before 2000 are not included in this instance due to the extremely low number of publications reported, which would have created a statistical artefact.)

separates priority areas within the tri-councils. Consequently, publications connected to the SSHRC focus on social science and arts and humanities. While we do not present the figures for the SSHRC separately for reasons of consistency, a closer look at the individual publications funded by SSHRC reveals that environmental and social sciences are strongly interlinked in Arctic science. The publications show great emphasis on the risks of climate change for regional and local living conditions and economic development. In all three countries, Arctic science, while remaining primarily focused on natural science, demonstrates that aspects of economic development, health research and social science are closely interlinked to climate and environmental changes in the Arctic.

In Canada, Arctic science publications have always covered a wide range of subject areas, but in the early 1990s the data show a greater focus on biochemistry, genetics and molecular biology (23.1% in 1990) and immunology and microbiology (7.7 in 1990). Since 1995, environmental science has appeared in the top three

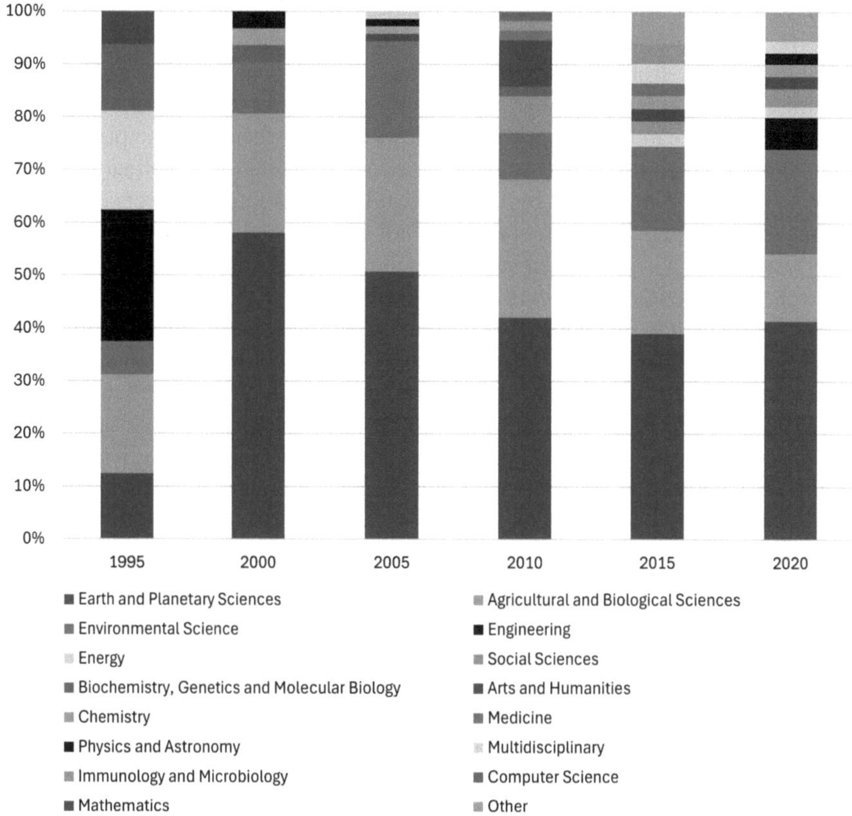

Fig. 3.6 Publication subject areas funded by the BMBF and DFG between 1995–2020. (Publications funded by the BMBF and DFG between 1990–2020. Documents by subject areas in SCOPUS.)

subject areas and has remained there since. It is also noticeable that publications in engineering accounted for 12.2% in 1995, then disappeared from the figures for 2000 and 2005, returning with 3.5% in 2010 and 4.1% in 2020. Since 2000, the share of social sciences has increased from around 6% to a share of 8.2% of publication in 2015. Arts and humanities research peaked in 2000 at 10.6% but subsequently decreased to 2.9% in 2020. As mentioned, over 90% of Arctic science is conducted in the HEI sector. This might be one reason why since 2015 multidisciplinary research seems to have become of increasing significance in Canada's Arctic science publications tally (2.3% in 2010 and 3.1% in 2020). In our understanding, this partly explains the variety of subject areas and the fluctuation of priorities. The practice of funding many shorter, university-based research projects also influences the variety of subject areas and reflects a fluctuation in priorities and pressing issues.

In comparison, publications funded by the RCN reveal that the top three areas account for around 70–80% of all publications. The lowest share of the top three was in 2020, with 60.4% of all publications. Social science publications consistently make up about 5% of all publications emerging from RCN-funded projects. Energy only appears in 2020 with 6.7%. It should be added that it is not possible to judge from the data to what degree individual publications represent multidisciplinary research. In our view this to some extent reflects the arrangement of Norwegian research funding, which follows a strong sectoral principle in which ministries fund their respective research areas. Multi- and transdisciplinary research projects may often be limited by funding regulations or larger projects dominated by one subject area. Furthermore, while not directly indicated by the data, the RCN's approach to 'Arctic science' differentiates between high north and polar research, a distinction that we will discuss further below.

For Germany, the first noticeable finding is that in the early 1990s, publications from various subject areas were already being recorded as Arctic science. Engineering was the most frequent subject area in publications in 1995, accounting for 25%, followed by energy and agricultural and biological sciences, each with 18%. Since 2000, the top three areas, earth and planetary sciences, agricultural and biological sciences and environmental science, have dominated German Arctic science publications. Standing at 94.4% of publications in 2005, their share decreased with the emergence of further subject areas such as art and humanities (8.8% in 2010) and social sciences (7% in 2010), as well as multidisciplinary science (3.7% in 2015; 2.3% in 2020). Considering that project-funded Arctic science only constitutes a small part of total Arctic science conducted particularly within the research associations, we expect the share of social sciences and humanities of both project-based and institutional funding to be significantly lower. The fact that Arctic science in Germany is embedded in a broader marine and polar research programme underscores the strong focus on the natural sciences, which include the areas of glaciology, permafrost research, as well as research in biochemistry, microbiology and molecular biology. Despite smaller publication outputs, Germany appears to have established itself as a significant contributor to international Arctic science due to its often more global perspective regarding the Arctic as part of marine ecosystems and the global climate.

Summarising these observations, we see similarities in the three countries regarding increases in both funding activities and the diversity of subject areas represented under Arctic science. Publication outputs show that natural sciences dominate Arctic science but are often, especially in Canada, interlinked with social sciences, health and humanities. Notably, there are different funding practices with regard to targeted sectors and funding allocation cycles. Canada focuses strongly on Arctic science in the HEI sector, while in Norway and Germany Arctic science is conducted to a large extent by research centres. The RCN has a distinctive funding cycle of 3–4 years, with higher budgets allocated to larger research initiatives and programmes. In comparison, Canada's research councils, like the BMBF and DFG in Germany, fund more projects with lower amounts reflecting more flexible allocation mechanisms.

3.2.3 Consultation Formats in Arctic Science/Politics Interfaces

On an international level, Arctic science has undergone significant changes since the 1990s, leading to the establishment of numerous international cooperation formats in science and policy-making. This has also increased awareness among scholars that previous approaches to research funding were 'particularly challenged by increasing developments towards more complicated and dynamic, border-crossing funding landscapes, where regional, national and supranational funders have proliferated from public, private and non-profit sectors' (Aagaard et al., 2021, 2). Specifically, as far as Arctic science is concerned, we have observed a diversification in two directions. On the one hand, there has been a growing emphasis on more inclusive knowledge co-production (Armitage et al., 2011; Berkes & Armitage, 2010; Nowotny et al., 2001; Yua et al., 2022) and community-based research frameworks (Doering et al., 2022; Ford et al., 2016), involving local and Indigenous communities. On the other, there has been an internationalisation or 'globalisation' of Arctic science, with an increasing number of states, international organisations and non-state actors participating in previously more closed national research funding systems (Sellheim & Menezes, 2022). The relationships and interactions between these organisations, however, are often shaped by the actors themselves in various formal and semi-formal ways. The role of different funding organisations and established funding practices support certain science/politics interfaces or 'coordination modes' (Lepori, 2011). In the following sections, we discuss different examples of science–politics consultation mechanisms including funding organisations from the three countries.

3.2.3.1 The Research Council of Norway as the Central Boundary Organisation

In Norway the RCN plays a key role in coordinating and advising research organisations and government bodies on Arctic and polar research. Although as the only national research council the RCN provides only around 25% of the public funding in Arctic and polar research, it is central in preparing Arctic science by coordinating and advising both research institutes and the government. The coordination of Arctic research between science and policy in Norway is characterised by 'public hierarchies' (Lepori, 2011) and sectoral principles, where the majority of public Arctic science is core-funded by government ministries. Norway does not have a national Arctic or polar research committee or network formats for polar researchers (EU-PolarNet 2 Consortium, 2022, 33). Instead, the RCN operates as a central boundary organisation (Guston, 1999) by stabilising the science/politics interface through its institutionalised structures of portfolio boards and internal processes of funding allocation. The RCN primarily collaborates with the Ministry of Education and Research, which specifically oversees the RCN's polar portfolio. Within the institutional landscape of public hierarchies, the initiative for consultations mostly originates from the ministries, usually addressing specific issues or problems. Regarding funding allocation, the RCN has its internal mechanisms. However, in semi-structured discussions with experts we learned that the RCN as a public agency must also consider government Arctic strategies and white papers when organising thematic calls, for example via the POLARPROG programme. Conversely, when ministries seek advice on drafting white papers and Arctic strategies, they turn to the RCN for scientific input. The RCN's advice is formally requested and provided in writing, and specific questions must be addressed by strict deadlines. As far as research centres like the Norwegian Polar Institute are concerned, the RCN has little direct influence on how research funds are spent or what research areas and priorities are planned.

The RCN is also responsible for the Svalbard Science Forum (SSF), which was established in 1998 to function as an attractive destination for polar and climate research and to improve research cooperation in Svalbard (RCN, 2025c). Several flagship programmes and interdisciplinary research consortia, such as the NANSEN Legacy, have been set up by various research institutions, and the RCN provides guidance in preparing and evaluating these long-term research programmes. These large-scale programmes are, however, directly funded by ministries. On the international level, the RCN both represents and coordinates Norwegian participation in international scientific cooperation formats. For example, it is part of the multilateral Arctic Science Funders Forum and appoints Norwegian researchers as representatives in the working groups of the International Arctic Science Committee (IASC). While the RCN is not the only connecting point between science and policy, it exemplifies how closely interlinked science and policy objectives are in Norwegian Arctic science, focusing on long-term monitoring tasks and interdisciplinary knowledge production for political decision-making.

3.2.3.2 Network Schemes as a Flexible Science/Policy Interface in Canada

Canadian Arctic expertise is widely decentralised at the national and regional levels across universities, colleges and non-profit organisations and government departments. Historically, Canada has lacked an effective consultation mechanism between Arctic science and politics at the national level to coordinate its funding and policymaking objectives (Lackenbauer, 2020, 80). Nevertheless, the federal research councils were established with a particular mandate to support post-secondary research and take on an advisory role to the Canadian government. The structural challenges of Arctic research coordination may partly explain why the tri-agencies launched initiatives such as the Networks of Centres of Excellence (NCE) early in 1989, aimed at funding 'partnerships between universities, industry, government and not-for-profit organizations' (Government of Canada, 2021). These initiatives provide opportunities to connect and coordinate research projects in a cohesive manner. Because Arctic expertise and special knowledge is widespread among scientific, Indigenous and local communities, various government agencies, and non-profit organisations, Canadian Arctic expertise and funding are well connected via organised transdisciplinary and cross-sectoral networks. ArcticNet is one of the largest and most important networks and embodies the objectives of transdisciplinary and international cooperation. ArcticNet encompasses organisations from all sectors, including federal and provincial government agencies, Inuit organisations, universities, companies and non-profit organisations. Furthermore, it comprises research teams from various countries around the world. By providing funds to the network, these resources can then be allocated to various projects and research initiatives within it. This allocation mechanism allows for responsiveness to special local needs and capabilities, making funding potentially more efficient.

This practice of developing network schemes has also been adopted by the tri-Councils, such as the NSERC Permafrost Partnership Network for Canada. While network schemes facilitate flexible and inclusive cooperation frameworks tailored to specific local or thematic needs, Benedetto Lepori points out that they might eventually lead to closed networks. This would make it challenging for small-scale or new researchers and groups to gain access. On the other hand, network schemes create spaces for interaction among different cultural and social values (Lepori, 2011, 361) and lower the threshold for scientific cooperation and knowledge co-production. The high number of funded projects relative to the funding amount, as observed in the funding trends, supports this argument. Focusing on Canada, Ibarguchi et al. (2018, 5) point out that '[s]upporting many projects at lower funding levels [...] may increase the type and diversity of research conducted simply because the door is open to more people' (ibid.). This practice could be explained by Canada's intention to 'facilitate research directed towards addressing Indigenous objectives' (ibid.).

For Canada, network schemes certainly provide a good science-policy practice for inclusive cooperation. However, these network schemes may only be a temporary solution to assist in establishing institutional and long-term sustainable structures at science/politics interfaces, so we see this as a still ongoing process. It was, for example, not until 2014 that the federal organisation Polar Knowledge Canada (POLAR) was established with the Canadian High Arctic Research Station (CHARS) to better coordinate and enhance the efficiency of research funding in Arctic and northern research, while also 'strengthening Canadian leadership in polar science and technology' (Government of Canada, 2025a). Whether this is indeed the latest step toward a sustainable institutionalised science/politics interface in Canadian Arctic science remains to be seen. To meet the varying territorial needs and research objectives, the funding of Arctic science in Canada is largely built on network schemes by the tri-councils, as well as on direct funds provided by provincial and territorial governments.[9] The still difficult and ambivalent relation between the federal government and the northern territories and its Indigenous populations (Bailes & Heininen, 2012, 58) may partly explain why an increasing number of

[9] Constitutionally, provinces have recognised legislative power. The three northern territories (Yukon, Northwest Territories and Nunavut) exercise power delegated by the federal government. (Government of Canada, 2025b).

inclusive network and collaboration formats have been set up to develop strategies and research objectives in more inclusive social processes of co-production. On the other hand, as relatively new large-scale Arctic science initiatives POLAR and CHARS indicate a stronger centralisation of science and policy-making in the Arctic. It remains to be seen whether, and how, this development will affect the decentralised landscape of expertise and the flexible network schemes in future Arctic science funding.

3.2.3.3 Formal and Semi-Formal Discussion Formats as Boundary Work in Germany

In Germany, we observe regular, but somewhat less formal coordination formats at the science/politics interface. Publicly funded polar research in Germany mostly developed around the AWI and the Federal Institute for Geoscience and Natural Resources (BGR) as the main research institutions. Arctic science at other research centres and universities very much depends on logistics and infrastructure operated by the AWI. In 1992, the German Research Foundation (DFG) established the National Committee SCAR/IASC as the central correspondence body for university-based polar research and funding agencies, including ministries. The AWI and the BGR are also represented in this forum, because universities rely on cooperation with these two research centres to have access to research stations, vessels and logistics for large-scale Arctic science. Also, the annual meeting is used to give updates on ongoing and planned research initiatives and programmes involving research centres and universities. This is important, for example, to coordinate applications and research time on the research vessel *Polarstern*. The AWI, as the main coordinating German polar research centre, also plays a key role in science–politics interaction. In 2013, the AWI established the Arctic Dialogue which is perhaps the most important coordination format in German Arctic research and policy-making today. The Arctic Dialogue was originally established to communicate AWI's research findings and knowledge to political actors and to strengthen the coordination between expertise and policy-making. It now meets every 6 months. Among its wide range of participants are various federal ministries, federal agencies, representatives of science, and non-governmental organisations.[10]

[10] Participants in the Arctic Dialogue are: the Federal Foreign Office, Federal Ministry of Research, Technology and Space, Federal Ministry of Agriculture, Food and Regional Identity, Federal Ministry for the Environment, Climate Action, Nature Conservation and Nuclear Safety, Federal Ministry of Transport, Federal Ministry of Defence, Federal Ministry of Economic Affairs and Energy, Federal Institute of Geosciences and Natural Resources, Federal Agency for Nature Conservation, Federal Environment Agency, Alfred-Wegener-Institute, Helmholtz Centre for Polar and Marine Research, German Council on Foreign Relations, German Institute of Development and Sustainability, German National Committee SCAR/IASC, German Aerospace Centre, Ecologic Institute, GEOMAR Helmholtz Centre for Ocean Research Kiel, GFZ Helmholtz Centre for Geosciences, Research Institute for Sustainability at GFZ, Potsdam Institute for Climate

The different consultation practices show that the interlinkage between science and policy in Arctic science is shaped by the actors involved. In Norway, the RCN is not only the central funding organisation but also relevant to science–policy coordination. Arctic science is, to a large extent, also Arctic policy-making, and coordination between the two is centralised in the RCN. In Canada, ArcticNet suggests a more holistic approach. The fact that over 90% of research funds go to HEIs underlines the significance of Arctic expertise and knowledge outside governmental organisations for policy-making. In Germany, the scientific community plays a central role in science–policy coordination. The AWI as the central German Arctic and polar research centre is essential for advising policy-making and also coordinating Arctic research at the science/politics interface.

To summarise, we have reconstructed the embeddedness of Arctic science and its funding in the research funding systems of Canada, Norway and Germany. By focusing on the central funding organisations that have financed Arctic science since the early 1990s, we have shown that governmental organisations are of central importance to Arctic science funding in all three countries. Ministries like the BMBF in Germany and research councils, such as the Research Council of Norway, or the NSERC and SSHRC in Canada, dominate the funding landscape. In Germany, however, the DFG, as a non-profit organisation, plays a significant role in university-based Arctic science and science–policy consultations. Judging by the funding inputs, Arctic science is primarily embedded in the institutional research sector in Germany and Norway, but in the HEI sector in Canada. All three countries share priorities in subject areas such as earth- and environment-related natural sciences. Nevertheless, there are differences regarding the variety and frequency of other subject areas funded as Arctic science. Publication trends suggest that Norway, Canada and Germany vary in their emphasis on subject areas connected to Arctic science, including, among others, social sciences, energy, engineering and medicine. Therefore, we suggest that, in addition to established organisational structures and practices in Arctic science funding, perceptions, values and experiences exert an influence in different funding systems. In the following thematic analysis of key strategic documents, we explore the similarities and differences in the motives and rationales for funding Arctic science and discuss the role of worldviews in the process of preparing Arctic science in the full science-policy circle.

3.3 Methodology

The following analysis of worldviews is based on strategy documents that address Arctic science and policies. Since we identified ministries or governmental agencies to be the main funding organisations, the key documents analysed here consist of

---

Impact Research, German Institute for International and Security Affairs, World Wide Fund for Nature (German Arctic Office, 2025).

national Arctic policy strategy papers as well as research agendas in which central objectives, research priorities and guidelines for funding practices for Arctic science are addressed (Table 3.1).

The material was approached on the basis of a mixed-method design. On the one hand, we developed code clusters based on the frequency of applicable terms and topics. On the other, we used thematic analysis (Braun & Clarke, 2006) to identify relevant themes in order to arrive at an in-depth understanding of particular perspectives and rationales connected to Arctic science and its role in different research funding systems. The aim of the thematic analysis was to capture, by observing the similarities and differences in the research funding systems and trends in Arctic science, the underlying scientific and policy-oriented rationales mirroring the varying funding practices and developments.

In a first step an initial coding process for the sections and paragraphs around the keywords 'research', 'science' and 'knowledge' was conducted to capture the variety of topics connected to research in the Arctic. Using a code co-occurrence analysis tool in the software atlas.ti, we identified the frequency of connections between initial codes and the keywords. As a result, we developed six broader code groups referring to (i) climate change, (ii) international cooperation, (iii) industry, economy and sustainable development, (iv) national research and policy-making, (v) security, and (vi) technology and resources.

In an explorative manner, we analysed the connections between themes in the code groups and key elements of worldviews. We reflected on the themes with respect to (1) geopolitical considerations and problems, (2) the prognosis for solutions, and (3) the epistemological foundations of how Arctic science is supposed to be conceived, and why it is necessary to fund it. The findings were then interpreted

Table 3.1 Strategic Arctic policy and research documents

Country	Key documents
Canada	The Northern Dimension of Canada's Foreign Policy Canada's Northern Strategy Our North, Our Heritage, Our Future Statement on Canada's Arctic Foreign Policy Canada's Arctic and Northern Policy Framework Polar Knowledge Canada Strategic Plan Polar Knowledge Canada Science and Technology Framework 2020–2025
Norway	Policy for Norwegian polar research 2010–2013 Research Strategy for the Arctic and Northern Areas The High North. Vision and Strategies Norwegian Polar Research 2014–2023 Polar Research Programme—POLARPROG (2015; 2018) Norwegian Arctic Policy (2014; 2017; 2020)
Germany	Rapid Climate Change in the Arctic: Polar Research as a Global Responsibility (2012; 2015) Germany's Arctic Policy Guidelines (2013; 2019) Polar Research Agenda 2030 Mare:N—Coastal, Marine and Polar Research for Sustainability

with regard to the worldviews described, and the reconstructed similarities and differences in Arctic science funding.

3.4 Worldviews in Preparing Arctic Science

'There are two kinds of Arctic problems, the imaginary and the real. Of the two, the imaginary are the more real' (Stefansson, 1945, 5).

This opening sentence in *The Arctic in Fact and Fable* by Vilhjalmur Stefansson (1945) encapsulates what we consider to be the main challenge when investigating funding priorities and trends in Arctic science. As described in the previous section, what is reported and funded *as* Arctic science does not follow a standardised formula. We therefore suggest that preparing science goes beyond examining the technical processes involved in funding within established organisational structures and practices. Understanding the preparation of science includes the social processes at the science/politics interface that determine what Arctic science means, what the driving issues are and what is expected from it. There are clearly observable concerns and interests across funding organisations in the three countries. Themes relating to climate change, unsurprisingly, feature as main subjects across all countries. However, other thematic areas, including engineering, energy and social sciences or health research, indicate variations in thematic priorities. In our thematic analysis of key strategic documents for Arctic science funding, we recognise that the perceptions of challenges and reasons for Arctic science are rarely consistent across countries and funding organisations. Instead, our analysis shows how key strategic documents issued by funding organisations reflect major science and policy concerns in specific national and local contexts. The different contextualisation is used to support and underline the necessity for research within and on the Arctic and its funding by the public.

The analysis shows that there are shared, explicit worldviews related to the Arctic. One is that of the Arctic as an endangered region in times of climate change; another dominant worldview conceives the Arctic as a resource frontier with various economic prospects, but also with risks. The analysis also shows that elements of different worldviews may interlink and overlap, creating more ambiguous, less explicit views. Key documents for Arctic science in Germany, Canada and Norway reveal similarities regarding the general idea of Arctic science and why it is important. However, the picture becomes more complex in direct comparisons of perceived geopolitical problems, prospects for possible solutions, and the epistemological foundations to justify publicly funded Arctic science. In the following section, we present worldviews in key strategic documents of Arctic science. We discuss how they are mirrored in the funding trends, organisations, and established practices in science–policy consultation formats. As a result, this chapter provides insights into how worldviews shape and reflect the dynamic process of preparing science at the beginning of the circle of knowledge production.

3.4.1 *The Changing Arctic: Science Between Climate Change and Economic Development*

Climate change is one of the most frequently addressed issues in Arctic science. In all key documents, the risks posed by climate change in the Arctic are explicitly referred to by the funding organisations. There exists a shared general perception of the Arctic as an endangered region. This diagnosis is summarised well, for example, in the polar research programme POLARPROG by the RCN, which states that climate change 'will have impacts on all ecosystem components' (RCN, 2015, 10). The Arctic research strategy by the BMBF (2012, 2015) outlines concerns about how the declining Greenland ice sheet affects global sea level rise and how diminishing sea ice cover alters the albedo effect and impacts marine ecosystems (BMBF, 2015, 6–7). Additionally, permafrost and gas hydrates are identified as 'unknown variables in the climate system' (BMBF, 2012, 8). The RCN emphasises in its polar research programme that the 'dynamic response of the cryosphere and the links between the cryosphere and the climate, including the impacts of climate changes on glaciers and permafrost thawing, are still not well understood' (RCN, 2015, 9).

However, climate change in the Arctic is also associated with prospects of economic development and resource exploitation, framing the region as a *resource frontier*. In Germany, we observe clear references to long-standing, and perhaps somewhat romanticised, images of the Arctic where 'the largely unknown regions in the Arctic harbour untapped natural resources' (BMBF, 2012, 5). According to this worldview, scientific efforts could help to better understand and assess the potential risks and opportunities associated with exploiting resources in the Arctic (BMBF, 2012, 11). By contrast, Norway presents a more rational image, acknowledging that the Arctic has:

> become more attractive as a region for natural resource recovery, in part because other resource-rich areas are characterised by political instability or uncertain framework conditions. Thus, global factors exert a significant impact on the economy, the environment and social development in the northern areas (RCN, 2011, 17).

From a Norwegian as well as from a Canadian perspective, the increased international attention to the Arctic evokes concerns regarding their perceived own Arctic sovereignty and leadership as Arctic nations. Given that perceptions of the Arctic as a potential *resource frontier* are linked to specific economic and national sovereignty concerns by Arctic states, the region is seen to have 'become an important crossroad where issues of climate change, international trade and global security meet' (CIRNAC, 2019, 4). On the one hand, changing climate conditions and melting ice covers on land and sea open new shipping routes and easier access to natural resources (ibid.). On the other, there are significant risks that '[c]ommunities in the polar regions will be affected to varying degrees by climate and environmental change' (RCN, 2015, 10). Norway emphasises the economic aspects and the impact of changing marine ecosystems on the availability of natural resources, such as fisheries and other 'commercial species' (RCN, 2015, 11). An increase in maritime activity in the Arctic Ocean might put additional pressure on the environment and already sensitive ecosystems (RCN, 2015, 11). As a non-Arctic state, Germany also highlights potential ecological risks associated with increased maritime activity and resource exploitation (BMBF, 2015, 11).

Together with growing attention to economic prospects, security concerns are also increasingly mentioned. It is important to highlight that security in the Arctic is often understood in a broad sense in terms of human safety and prosperity, for example in the northern parts of Canada (DFAIT, 2000, 2). The Science and Technology Framework of POLAR Knowledge Canada states that 'Northerners [meaning people living in Canadian northern territories including Indigenous peoples] could benefit from a better understanding of how changes in the environment affect food security, community well-being and traditional lifestyles' (POLAR, 2020b, 18). The security aspects associated with economic issues also include sustainable development, food security (CIRNAC, 2019, 5) and energy access (CIRNAC, 2019, 44). Such a broad understanding also includes so-called 'hard' issues. In Canada's understanding, 'safety, security and defence are essential prerequisites for healthy communities, strong economies and a sustainable environment' (CIRNAC, 2019, 50). Norway also asserts that the concept of security is 'not only a matter of military or defence security' (RCN, 2011, 13), but includes the ability to respond to changes in economic, scientific, and political activities in the region (RCN, 2011, 13). Both Norway and Canada view security in the Arctic as an international challenge, but also as their special national responsibility as Arctic states. For Norway, Svalbard and the surrounding sea areas where it exercises its sovereignty are of geopolitical significance (RCN, 2011, 6). Similarly, Canada emphasises its commitment to 'exercising the full extent of its sovereignty, sovereign rights and jurisdiction in the Arctic' (Global Affairs Canada, 2010, 5). Germany acknowledges that coastal Arctic states have rights under international law that allow them to regulate maritime research in their territories and economic zones. However, the government's Arctic policy guidelines emphasise that these states 'are also under an obligation to foster international cooperation' (Auswärtiges Amt, 2013, 9, 2019) within these areas. Naturally, the documents analysed do not explicitly point out that there are conflicting or contradicting diagnoses. The necessity for international cooperation and interdisciplinary research to enable a better understanding of climate change and the economic risks in the Arctic is mutually shared and supported. So, the documents subtly highlight different aspects of the issue on global, national and regional scales with varying policy implications. Canada and Norway see themselves as having special national responsibilities to regulate and control what is considered their own territory in the Arctic, while Germany also emphasises the necessity of keeping the Arctic open and accessible for international scientific activity.

Thus, there is far from a consensus on what is considered Arctic and what the pressing issues and geopolitical concerns are. It depends on the specific political contexts and the different actor constellations (Wilson Rowe, 2018, 6). The examples of funding organisations in this chapter show that perceptions of geopolitical challenges and the institutional framework conditions for public research funding influence one another. The analysis does not allow us to reach conclusions on specific directions of influence. Nevertheless, it does underline that geopolitical perceptions and the policy dimensions of Arctic science vary across research funding systems and funding organisations. This supports our assumption that preparing science through research funding in the context of the Arctic is more than a technical process of funding allocation and evaluation. It shapes and mirrors specific geopolitical perceptions and the necessity for scientific activity.

3.4.2 International Cooperation and National Funding Structures

The common understanding of an *endangered Arctic* is closely linked to the need for international scientific cooperation as a prospective solution. The documents we analysed highlight the international dimension of Arctic science as a global responsibility in the context of climate change. The primary approaches to achieving international collaboration involve the interdisciplinary design of Arctic science as far as its content and thematic priorities are concerned, as well as the establishment of shared infrastructure and logistics for collaborative practices. For example, changes in the climate are recognised as occurring earlier or more significantly in the Arctic and thus 'can provide an early warning of change' (RCN, 2010, 15). This is particularly important for '[d]eveloping an Earth system science approach using models that link regional studies in the polar regions with global studies' (RCN, 2010, 17) and the 'development of global climate patterns' (BMBF, 2012, 5). Establishing and enhancing global climate models through interdisciplinary Arctic science is seen as a means to better understand changes in the Arctic and their impact on global climate systems. Nonetheless, Arctic science is a complex and costly endeavour.

The second reason for increasing international cooperation is therefore to be found in shared infrastructure and logistics. In Germany, Arctic science is associated with high standards for the infrastructure and equipment made available for international cooperation, thus ensuring Germany's international significance in the field. Norway views the establishment of advanced research infrastructure on Svalbard as a unique and special platform for international collaboration (RCN, 2010, 6). The funding trends mentioned above underline the shared perceptions and priority settings for international Arctic science in basic natural sciences, as well as in the development and engineering of infrastructure. Moreover, this includes various items of hardware and software for modelling climate developments to better understand the Arctic within a complex global climate system.

However, as observed, research funding for Arctic science is still largely allocated at the national level, where the varying regulations governing allocation processes pose challenges to coordinated international Arctic science. While on a national level project funds can be allocated relatively quickly, international Arctic science projects require early and long-term preparation. Coordinating different funding cycles and regulations is a major undertaking. Consequently, Arctic science, whether conducted on a national or international level, is predominantly funded nationally and is often linked to policy-driven objectives. While the worldview of an *endangered Arctic* seems to entail the promotion of international scientific cooperation as a means of generating knowledge and finding solutions, the actual funding structures and established practices that are crucial to preparing science appear to be based on an understanding of Arctic science as a national endeavour.

As illustrated by the formats of science/politics interfaces described above, coordination between science and policy has to mediate between an internationally oriented scientific community and a more nationally oriented policy-making process. Germany, Canada and Norway have different approaches to overcoming this tension through various formal and informal consultation mechanisms. The establishment of the RCN as the only central research council and boundary organisation in Norwegian Arctic science suggests the presence of strong public hierarchies in, and a policy-oriented basis for, Arctic science funding (Lepori, 2011). In Germany, consultation formats are less formalised. The Arctic Dialogue, organised by the Arctic Office of the AWI, as well as the National Committee SCAR/IASC coordinated by the DFG, provide fora for information exchange and the discussion of the general situation and future prospects of Arctic science. The National Committee provides an important platform to coordinate future research projects among HEIs and the research centres organised in research associations. In light of the limited nature of resources, the National Committee provides a communication channel between science and policy and coordinates future funding calls by the BMBF, identifying overlaps in future research plans that can be merged. In Canada the NSERC and the SSHRC, as central research councils, have established various instruments to coordinate widespread networks of HEIs, local and Indigenous communities, and federal as well as territorial government agencies. In Norway and Germany, Arctic science funding primarily supports research centres and the implementation of research programmes by mean of a consortium or research association, while Canada exhibits a more holistic approach in coordinating between HEIs, research centres, policy and society.

3.4.3 Understanding the Arctic: Values and Epistemological Foundations

Our analysis shows that (1) *the endangered Arctic* in the context of climate change, as well as (2) *the Arctic as a resource frontier*, focusing on the exploitation and management of natural resources, constitute two commonly shared and explicit worldviews in preparing Arctic science. However, other images, values and rationales overlap with them: Canada places particular emphasis on the economic and social consequences for local and Indigenous communities in its northern territories, while Norway associates climate change with the responsibility of scientific activities to foster the development of a sustainable economy and management of natural resources. Both Canada and Norway also indicate that *national Arctic science* worldviews are integral to exercising their sovereignty and authority as Arctic states. This is often justified by referring to a historical and cultural heritage. References in research strategies to the importance of the co-production of knowledge with Indigenous knowledge-holders in the northern parts of the country also plays an important role for legitimising research funding, particularly in Canada. For Germany, the shared responsibility of understanding the Arctic as part of a global climate system, along with its efforts to provide advanced technology and scientific expertise, encompasses views of a *global Arctic*. However, Germany as well needs to take certain perceptions from Canada and Norway into account, notably also in order to ensure access to research areas on Indigenous homeland. Thus, Canada, Norway and Germany utilise various arguments and rationales to legitimise their efforts in Arctic science. Its role appears to depend on the issues identified as problematic, and on whether the prospects of solutions are viewed as lying more in a national or an international context (Engels, 2005, 10). Both the worldview of an *endangered Arctic* as well as the worldview of *the Arctic as a resource frontier* can be combined with different values. In addition, and depending on the intervening elements of other worldviews, the endangered Arctic can have specific national and local aspects. Previous studies on science–politics relations in Arctic science emphasise that the significance and purpose of science in policy-making is a 'complex and multidimensional issue pervaded by divergent and contradictory views' (Łuszczuk et al., 2020, 609). Thus, worldviews are rarely consistent but are composed of many perspectives. In the context of Arctic science funding in nationally organised funding systems, a general understanding of the Arctic as situated in a changing climate allows for communication and cooperation at the international level. However, to legitimise publicly funded science, the general diagnosis of an *endangered Arctic* needs a specific contextualisation by funding organisations in order to underline the necessity of scientific evidence and knowledge. The Canadian and Norwegian funding systems for Arctic science exemplify a nuanced understanding of the Arctic and its significance in an international or local context. This is particularly evident in comparison to Germany. Worldviews provide idealised frameworks for defining the Arctic. This idealisation seems to be an important foundation for the legitimacy of public research funding.

3.4.4 Defining Arctic Science

One notable finding reveals how Canada and Norway use different terms and frames to highlight the purpose of Arctic science at both national and international level. Since 'Arctic science' is not an established term, its usage is often directly linked to the political perception of the Arctic as a distinct region. This is particularly evident in scientific and political arguments for funding Arctic science put forward by Arctic states. The first step in legitimising Arctic science activities is to specify *what* Arctic science entails and where it occurs. Canada and Norway draw distinctions between domestic Arctic regions, such as the High North in Norway (orig. *Nordområdene*; Skagestad, 2010; Norwegian Ministry of Foreign Affairs, 2011; Østhagen, 2021) or the northern dimension of the nation policy in Canada (DFAIT, 2000; INAC, 2009). As early as 2000, the Canadian government expressed concerns about the geopolitical implications of a 'globalised' Arctic science and policy, which 'exposes all regions to new political, economic, social and environmental forces, which often diminish regional control over events' (DFAIT, 2000, 4). From the perspective of Canada and Norway, their status as *being in*, or *being part of* the Arctic is essential. In this context, Norwegian knowledge in the High North is not the same as Canadian knowledge in northern research, as they refer to different parts of what is generally labelled as the Arctic. For preparing science this means that the Research Council of Norway has established two separate Arctic research programmes based on the geographic demarcation: A polar research programme highlighting Arctic science as part of a globally connected *endangered Arctic* (RCN, 2018, 7), and a High North research programme referring to the Arctic as a *resource frontier.* Norway as an Arctic nation has specific interests in managing and protecting natural resources and ecosystems in its Arctic territories including Svalbard, and in exercising its national sovereignty (RCN, 2011, 11). In Canada, the expressed need to better understand the 'changing Arctic' (Steinberg et al., 2018) also refers to requirements of targeted knowledge-production on local, national and international levels. Canada's Arctic research funding attaches particular importance to research in which 'Indigenous knowledge and scientific findings create new knowledge that is made available to all Canadians for use in policy and decision making, and tailor knowledge products to different audiences' (POLAR, 2020b, 15). For both Canada and Norway elements of *national Arctic* science worldviews not only refer to intrinsic convictions of having special experience and connections by *being* in the regions, but also to special knowledge needs. They also express sovereignty and responsibilities, thus drawing geopolitical distinctions between a *national* and a *global Arctic*.

Germany's approach to Arctic science needs to be understood differently. The Federal Ministry of Education and Research (BMBF) released two strategy papers both entitled *Rapid Changes in the Arctic: Polar Research as a Global Responsibility* (BMBF, 2012, 2015). These documents particularly outline the importance of fostering science in and on the Arctic as a global task and responsibility within worldviews of the *endangered Arctic*. Perspectives for German Arctic research are generally outlined within a broader understanding of polar research (Heinemann et al., 2017). The Polar Research Agenda 2030 (orig. *Polarforschungsagenda 2030*) of the National Committee SCAR/IASC of the German Research Foundation (DFG) illustrates how the scientific community rarely speaks of Arctic science specifically, but understands it as part of polar science, in order to emphasise its connectivity within a global climate system. For Germany, being part of a *global Arctic* carries legitimacy and, in light of climate change, a shared responsibility to conduct Arctic science. In this sense, Germany acknowledges that knowledge about the Arctic can best be gained by being present in the region. Without direct geographical connection, it relies on effective international cooperation and the efficiency and high quality of its research infrastructure, which ensures access to the Arctic. Despite shared general worldviews, the subtle differences in the argumentation and legitimisation of Arctic science being funded and eventually conducted are diverse and multidimensional. Within the shared worldviews behind Arctic science, we find different diagnoses, prospects of solutions and value-related foundations indicating the involvement of a number of different justifications for Arctic science and its preparation.

3.5 Conclusion

In this chapter, we have examined the central funding organisations and funding practices that supported Arctic science in Canada, Norway and Germany from 1990 to 2020. Using the RCN (Norway), NSERC and SSHRC (Canada), and BMBF and DFG (Germany) as examples, we have reconstructed the role of funding organisations as central actors in Arctic science and policy-making systems, as well as funding trends, on the basis of financial inputs and publication outputs over the same period. We have highlighted specific institutional framework conditions and funding practices. Through analysing the underlying rationales and worldviews expressed in central Arctic science strategies, our aim has been to provide a better understanding of the similarities and differences in approaches to Arctic science funding within the science–policy circle.

Canada, Germany and Norway exhibit different emphases in their funding practices. Canada focuses on funding research in the HEI sector and the use of network schemes to account for particular local circumstances and to enhance knowledge co-production across scientific and Indigenous knowledge systems. Norway and Germany primarily fund Arctic science in research centres outside HEIs with targeted, programme-oriented means. In addition, they also support research projects and large research consortia with the aim of connecting research centres and HEI. The frequency of funding allocations also varies, with the RCN having distinctive three- to four-year cycles, providing larger amounts to fewer long-term projects, whereas amounts in Canada and Germany tend to be smaller, but support a larger number of projects. Furthermore, the NSERC and SSHRC are separate research councils funding specifically Arctic natural or social sciences in Canada. Collective tri-council funding initiatives like the Networks of Centres of Excellence (NCE) play a significant role in Canadian Arctic science, as the case of ArcticNet illustrates. By contrast, Norway's RCN and the BMBF and DFG in Germany fund all areas of research. Norway, however, distinguishes between polar and high north research, while Germany has its Arctic science embedded in broader polar and marine research programmes. A clear distinction of funds for 'Arctic' science therefore significantly depends on the definition of research interests.

Regarding funding trends, this chapter has also traced how funding for Arctic science has increased significantly over time, indicated by research output of publications. The data collected from the Scopus database indicate that Arctic science remains primarily oriented towards the natural sciences (environmental sciences, Earth, planetary and biological sciences) in all three countries. Nevertheless, Arctic science also encompasses other subject areas that vary over time revealing changing trends and priority settings. For example, in Germany engineering and energy topics were more present in Arctic science publications of the early 1990s. Since the 2000s, Arctic science in Germany has been dominated by natural science research, particularly in connection with the marine environment. In Norway since 2010 an increasing number of research projects from the social sciences and the humanities, as well as from engineering and energy-related fields, appear to have been funded, albeit trends are somewhat difficult to monitor since the data do not distinguish between polar and high north research, where the trends might show distinctive differences. In Canada we have recognised an increase in social Arctic science and multidisciplinary sciences. In combination with the distinctive funding practices in the three countries, we see similarities in the overall priority areas in Arctic science, but significant differences in terms of who is funded, and for how long.

To better understand the similarities and differences, we have analysed the main worldviews in key strategic Arctic science documents provided by the funding organisations. We found indications of explicit worldviews shared by all three countries, along with different values and interests from various perspectives, resulting in more nuanced and ambiguous worldviews in specific political or cultural contexts. We also found that overlapping ideas and values in different Arctic strategies address what we have called, following Stefansson (1945), different *'imagined'* problems highlighting distinct aspects of the *'real'* complex and multidimensional problems connected with the Arctic (Stefansson, 1945). As a result, this chapter shows that preparing Arctic science through funding is not only a technical procedure, but needs to be understood also as a dynamic social process within varying institutional landscapes and established practices in research funding systems. Worldviews can provide important ideal frameworks in these processes to bridge different perceptions in science and policy and different countries. Furthermore, worldviews help to define boundaries between scientific knowledge production and policy-driven decision-making. We have identified two specific worldviews shared by all three countries. First, the *endangered Arctic* emphasises the challenge of climate change and the need to better understand its development through international cooperation as well as through special local and regional cooperation formats. Second, the Arctic is viewed as a *resource frontier* with various economic prospects of resource exploitation but also risks, including to territorial sovereignty and free access for international science. However, these shared worldviews are also reinterpreted in national or global contexts through various intervening ideas and beliefs leading to worldviews that we have summarised as more *national Arctic*-oriented and, in contrast, more *global Arctic*-oriented. Needless to say, the finding that Norway and Canada as Arctic states refer more to *national Arctic*, with Germany referring more to *global Arctic*, worldviews, does not come as a surprise. The main point of the present chapter was to show not that these differences exist, but how they are also reflected in funding practices and in the way that Arctic science is embedded in public research funding systems. All in all, these worldviews are not

Table 3.2 Worldviews in Chap. 3

Worldviews of Ice	Case study 1: 'Preparing Science'
Endangered Arctic	
Diagnosis	Climate change
Solution	More interdisciplinary and cross-sectoral research
Epistemological foundation	Knowledge through international cooperation and local co-production
Arctic as a resource frontier	
Diagnosis	Contest for resources and ecological risks
Solution	Combination of science and economic development
Epistemological foundation	Science for the benefit of progress and applicability for sustainable development
National Arctic science	
Diagnosis	Arctic sovereignty
Solution	Focus on 'own' Arctic
Epistemological foundation	Knowledge by 'being' in the Arctic
Global Arctic	
Diagnosis	Shared responsibility
Solution	Showing presence around the circumpolar region
Epistemological foundation	Knowledge through exchange of information

mere projections of the geopolitical positions and strategic interests held by influential actors. Instead, they emerge out of and also shape complex institutional and organisational constellations in funding systems, permeate the ways science and politics interact—and lay down the very foundations of what Arctic science actually is and why it matters for societies (see Table 3.2 for a summary).

Chapter 4
Constructing the Arctic II: 'Doing Science'

Abstract This chapter analyses those actually 'doing science' in the fields of permafrost and glaciological research. We are interested in the extent to which specific worldviews are present in Arctic scientific communities. Interviews with researchers paint a broad picture of (world)views of and on the Arctic, asking how, for example, the Arctic is understood and framed as an object of research, how evolving Arctic research priorities are seen, how scientists themselves consider their relation to politics, etc. The chapter presents a multi-faceted picture of natural science in the Arctic, and in particular also of scientists themselves, but identifies a range of outstanding common threads when it comes to self-understandings as scientists as well as the relation between politics and science.

Keywords Permafrost research · Glaciology · Indigenous knowledge

4.1 'Doing Science': The Case of Permafrost Research and Glaciology in the Arctic

In order to unravel the circular relation between science and politics in its various forms, the present chapter takes a closer look at the individual level, that of those who are actually 'doing' research. Here, as already noted in the introduction, the *worldviews of ice* are located in the attitudes and views of researchers who work with ice in a narrow sense, that is permafrost researchers and glaciologists. This focus on ice results from the fact that the most prominent representations of the Arctic depict it as a region mostly covered by snow and ice, but rapidly losing precisely that characteristic due to climate change. A melting Arctic opens up a range of imaginaries regarding the region's present and future: as a tourist destination, as a repository of vast natural resources, or as a transit area for faster shipping routes between the Atlantic and Pacific Oceans. Thawing permafrost and the mass loss of glaciers are among the main direct sources of these developments, in addition to the disappearance of sea ice. We focus on glaciology and permafrost research as areas covering the first of these subject areas, and thus on the terrestrial realm (with the

M. Albert et al., *The Worldviews of Ice*, Springer Polar Sciences,
https://doi.org/10.1007/978-3-032-25945-5_4

exception of coastal subsea permafrost research pertaining to the maritime realm). Terrestrial fieldwork is subject to national and regional permitting and operates out of research stations or field camps. In line with the analytical focus of the previous chapter, we concentrate on researchers affiliated with Canadian, German and Norwegian institutions.

At the core of our research lies an interest in how worldviews come into being through a circular process that includes funding, the actual doing of science and the interaction between science and politics in forums such as the Arctic Council. In this sense, we are not only asking what the worldviews held by permafrost researchers and glaciologists in the Arctic 'are', but also how they inform their daily research practices. The present chapter first provides some context for the research fields we examined, permafrost research and glaciology (Sect. 4.2), followed by an explanation of our methodology (Sect. 4.3). Building on this, we then analyse researchers' worldviews, deliberately focusing on their function in a wider social and political field (Sect. 4.4).

In conclusion, we aim to demonstrate that 'doing' permafrost research and glaciology in the Arctic is far from being a 'pure' scientific and value-neutral exercise that is then at some point confronted with the political. We show that researchers' worldviews matter in shaping views of what the Arctic and Arctic research *are* and *what they should be*. We find that these views become particularly apparent when considering (1) the reasoning behind research topics and knowledge gaps, (2) researchers' reflections on people's own career choices, and (3) their direct engagement with media, policy-makers and the public. After all, when it is up to researchers to decide which perception of the Arctic is to be shared, and what role to ascribe to themselves in this story, we detect a rather strong line being drawn between science and politics, based on a narrow understanding of their role and an emphasis on the Arctic as an endangered region. In this respect, worldviews fulfil an ordering function and provide orientation in times of uncertainty. While, on the one hand, research into an *endangered Arctic* could possibly be more easily justified by reference to this danger, we demonstrate how, on the other hand, such a worldview may also become an obstacle, for instance when it comes to deciding on research priorities between basic and applied science. We conclude that these worldviews and role perceptions work in a reciprocal way, with researchers incorporating certain expectations and role perceptions while at the same time being the subject of others' expectations and projections.

4.2 Context: Getting to Know the Research Field

Situated at the intersection of critical geopolitics and science studies, we look at *worldviews of ice* that are—consciously or unconsciously—present in the research practices of natural scientists. We draw on the research tradition of critical geopolitics in order to acknowledge that different 'Arctics' exist, and are constructed to serve different purposes. We also draw on science and technology studies (STS) in order to reflect how, in modern society, evidence-based policy relies on science and technology, and therefore science itself becomes a sociological subject (Maasen et al., 2012, 10). With regard to evidence-based policy we stress that the epistemology itself should not be taken for granted. In contrast, the notion of something being 'evidence-based' presupposes the assumption—or worldview—that presents science as being objective, verified and validated to inform politics, thus describing a somewhat linear relationship between science and politics.

Building on work such as Richard Powell's *Studying Arctic Fields* (2017) or Jessica O'Reilly's *Technocratic Antarctic* (2017), we arrive at a better understanding of research practices in the Arctic by inquiring into the role that worldviews as well as very individual (role) perceptions play in this context. According to Frederick Grinnell, we understand science 'as something that people do' (Grinnell, 1992, 131). Science is never conducted in a vacuum, but always within specific (for example, national, societal or institutional) environments. This implies that factors such as age, gender, nationality, socio-economic background, socialisation, career stage, general job situation, institutional structures or 'Arctic experience' can all play a role in the doing of science. Studying scientists who are part of an international research community implies an understanding of the function of epistemic communities (Haas, 1992). Especially with regard to the development and negotiation of different worldviews within these communities, we consider role perceptions, notions of expertise and scientific authority as well as related power-exercising functions to be indispensable for our study.

Needless to say, we are aware of the fact that complex research questions require trans- if not interdisciplinary approaches, and therefore also consider multifield research. Glaciology and permafrost research draws on a wide range of natural science disciplines, and scientists are trained and educated in geography, geology, chemistry, physics, biology, ecology and more. Both research fields are rapidly advancing with the development of new technologies, such as satellite imagery and computer modelling. Likewise, 'the Arctic' as such is rarely studied as a single research object in the natural sciences. Arctic regions are diverse in terms of their landscape, climate, ecology and habitat, let alone their societal, economic and political characteristics. It must therefore be said that, while we are interested in how *worldviews* fulfil an ordering function and help overcome uncertainty while leading to common expectations of *the Arctic*, we are to some extent confronted with the reality of fieldwork, that is, the impressions, experiences and information of an individual scientist about very specific places in the Arctic. The scientists' self-descriptions are very revealing in this sense, because they mirror this diversity in the ways in which the *doing of science* could be described. A permafrost researcher, for instance, who works exclusively in the Arctic describes himself as a polar researcher for a public audience. A glaciologist would introduce herself to a natural science community as a polar researcher, when her polar expertise is required, but elsewhere call herself a glaciologist. Another glaciologist finds that polar researcher reminds them of the 'polar hero narrative', which sounds cool. In fact, the self-descriptions show a differentiation depending on the targeted audience—from media attributions to notions within one's own research community to public appearances, thus reinforcing ideas and images of the research field and the Arctic in general.

4.2.1 Permafrost

Permafrost is defined as any subsurface that has continuous temperatures of 0 °C or lower for two consecutive years. Arctic regions contain most of the world's permafrost, 'where it underlies about 15 per cent of the exposed land area' (Westerveld et al., 2023, 13). In the Earth's geological past, different climatic conditions have prevailed since the end of the last ice age 12,000 years ago: cold glacial periods and warmer interglacial periods, with the latter representing our present situation. Most of the terrestrial permafrost we experience today was formed towards the end of the last ice age in regions which were not covered by glaciers (Westerveld et al., 2023, 17). Scientists distinguish between continuous (90% of the area), discontinuous (50–90%) and sporadic (10–50%) permafrost, and the thickness of frozen soil layers can vary between less than one and 100 m. The largest area of permafrost globally can be found in Russia, followed by Canada, with nearly half of Canadian territory covered by frozen ground (Gruber et al., 2023). In these regions, scientists are also studying subsea and coastal permafrost. Permafrost in mainland Norway can be found in the northern provinces Nordland, Troms and Finnmark. In Southern Norway, it exists on bedrock slopes and has in the last years experienced more scientific attention, especially related to risk management (e.g. rockfalls, rock–ice avalanches). This contrasts with the extensive research and long-term monitoring of permafrost in Svalbard (Westerveld et al., 2023, 38). In Germany, a few mountain permafrost deposits can be found for instance on the Zugspitze, the country's highest mountain.

Thawing permafrost in Arctic regions has several local and global implications. Estimates suggest that around five million people live on permafrost today, distributed across 1162 Arctic and sub-Arctic communities (Westerveld et al., 2023, 28). Once the frozen ground is no longer solid, infrastructure, such as housing, highways and roads, pipelines or airfields, is at risk of collapse. With less sea ice, coastal erosion is advancing, posing a serious risk to people living close to the sea and eventually resulting in the relocation of communities. Seen from the air, many permafrost landscapes feature polygonal structures. If the permafrost is particularly ice-rich, the ground surface sinks due to the volume loss of the ice during thawing, creating characteristic thaw lakes. This process is also described as thermokarst. Permafrost provides insights into the history of our planet. Rocks, vegetation and animals that lived in the non-glaciated areas in the short summers during the last ice age were exposed to the cold without the insulating glacier ice and have been preserved for centuries as if in a freezer. Due to the cold, this organic material has not decomposed, and so the permafrost today contains enormous amounts of carbon. A warming Arctic increases the temperature in the active layer of the permafrost, that is, the surface layer that thaws and freezes as seasons change. Put simply, once this active layer gets thicker and warmer, the organisms in the soil become active and start decomposing the organic material. Because of the complexities of this carbon cycle, researchers face huge uncertainties when it comes to estimating the amount of carbon dioxide or methane that will escape from the thawing permafrost and be released into the atmosphere (Chadburn et al., 2017). Furthermore, not only carbon or methane, but other pollutants such as radioactive waste, mercury, heavy metals or even viruses and bacteria from carcasses stored in the now thawing permafrost could pose a potential environmental hazard (Langer et al., 2023).

4.2.2 Glaciology

Unlike permafrost, which is mostly hidden from human eyes under the active soil layer, glaciers truly possess picture power. More than 200,000 glaciers exist on Earth (Hagg, 2022, 62). A glacier is defined as a massive accumulation of ice formed from snow, firn and glacial ice as well as meltwater and entrained rock, which actively flows downhill due to its large mass (Hagg, 2022, 3). In Canada, glaciers exist in western and northern mountain ranges, such as the Rocky Mountains and the St Elias Mountains—the largest subpolar icefield which stretches from Yukon territory to Alaska, USA. Most of the country's glaciers, however, are located in its Arctic Archipelago. Norway counts a number of mainland mountain glaciers—like, for instance, Jostedalsbreen in southwest Norway, which is the largest glacier on the European mainland. These glaciers are tourist destinations and important energy resources for hydrology. On Svalbard, in Arctic climate conditions, approximately 60% of the land is covered by glaciers (Nuth et al., 2013). In Germany, only four glaciers are left in the Alps and they are expected to vanish at some time between 2030 and, at the latest, 2050 (Mayer et al., 2021, 44).

Scientists distinguish between different types of glaciers. Greenland and Antarctica are today's only ice sheets, that is a glacial area resting on land extending more than 50,000 square kilometres. The melting of ice sheets contributes to sea level rise. Frequently cited numbers estimate a global sea level rise of 58 m following a complete melt-down of Antarctica (The IMBIE team, 2018), and seven meters if the same thing happened in Greenland (Morlighem et al., 2017). Ice caps and mountain glaciers are smaller remnants of ice sheets that existed in the past. Many of them grouped together in one region create ice fields. Tidewater glaciers, at their ablation zone, terminate in the sea—either through fiords or ice shelves—and eventually break-off or calve in a spectacular manner.

Research on glacier mechanics and hydrology seeks to understand how they flow, how they have changed and formed the landscape in the past (geomorphology), and how they will change the landscape in the future. In addition, ice cores taken from glaciers provide a unique climate archive. Air bubbles trapped in the ice cores can, depending on location, range in depth from a few to several hundred meters, allowing us to reconstruct atmospheric features of the past.

4.3 Methodology

At the heart of this chapter are semi-structured interviews, conducted with permafrost researchers and glaciologists working at Canadian, German and Norwegian institutes between 2022 and 2024. In summer 2022, a test phase was carried out, in which the questionnaire was discussed with two scientists, after which no further changes were necessary. While, where possible, in-person interviews were preferred, for practical reasons half of the interviews took place online. In some cases, they were conducted by two interviewers from the project team, but mostly one-to-one. All interviewees were informed about the project's objectives and motivation beforehand. The length of the interviews varied from 60 to 90 min. Further, all interviewees agreed to their interviews being recorded and a cleaned transcript of the recording being generated for internal analysis. All in all, we were able to talk to 21 researchers—from PhD students through mid-career and post-doc to senior/professor level. For reasons of anonymity, the affiliations are not disclosed.

In order to identify interview contacts, we used a snowball sampling approach. Already established contacts pointed us to their colleagues and people who were interested in our project as well.[1] Having researched and knowing the central research groups, institutes and university departments for permafrost and glaciology in all three countries, we created a list of people to talk to and approached them by email. In addition, the growing networks of international contacts made over the course of the project at conferences such as the Arctic Circle Assembly, Arctic Science Summit Week, Arctic Frontiers and Arctic Congress enriched our portfolio. Further, participatory observations and informal discussions provided us with helpful insights into knowledge production in the Arctic. We can highlight reflections on the status of our research through poster presentations and participation at two central events for the research communities: the IASC Network on Arctic Glaciology Annual Meeting 2024 in Obergurgl, Austria and the International Conference on Permafrost (ICOP) 2024 in Whitehorse, Yukon, Canada.

Our questionnaire covered six major topics:

- *The Arctic as a research object*; including questions about individual motivation, research interest, and fieldwork experiences.
- *Arctic research priorities*; including questions about funding, differences between national and international research agendas, and involvement in research planning processes.
- *'Multiple Arctics'*; including questions about regional differences in the Arctic and media coverage of the Arctic and Arctic research itself.
- *Exceptional Arctic?—Arctic research cooperation(s);* including questions about self-identification within the Arctic research community, the characteristics and

[1] Having said this, we are well aware of the bias created through the fact that researchers who are more reflective about their research might also be more open towards our project. We address this limitation by contextualising the empirical material within the broader context of Arctic research, including information from case studies one and three.

challenges of international scientific cooperation, the impacts of the war in Ukraine and of the COVID-19 pandemic on research.
- *Arctic vs. non-Arctic states;* including questions about the different role perceptions of Canadian, German and Norwegian Arctic research.
- *The role of science in politics and society;* including questions about science communication, and scientists' responsibilities.

Knowledge production in the Arctic includes more than a Western scientific understanding of permafrost and glaciers. However, an in-depth discussion of the role of Indigenous and local knowledge, colonial past and presence, experiences with research collaborations, as well as perceived possibilities and limits of braiding different knowledge systems exceeds the scope of this chapter.

We used atlas.ti software to conduct a qualitative content analysis of the interviews. Corresponding to the strong exploratory character of our study, we approach the material in different ways throughout this chapter, as well as throughout the book. The initial approach to worldviews was literature-based (see Chap. 1), and the guiding categories as well as the questionnaire structure were arrived at in a deductive fashion. The coding and analysis of data, however, followed an inductive approach, though it was oriented towards these predefined thematic blocks. Central arguments were extracted from the interviews and will be discussed and contextualised with regard to their various facets.

4.4 Worldviews of Permafrost Researchers and Glaciologists

Having approached the *worldviews of ice* from a rather theoretical angle at the beginning of this book, this section now identifies worldviews and their functions in an empirical and interpretative way. While there are many interesting aspects to consider following the extensive literature review, and while some worldviews are more strongly represented and others more hidden, four focal points emerged that ultimately provided the structure for the empirical material. Thus, as the analysis progresses, four key worldviews will be unravelled (the *endangered Arctic, unknown Arctic, Arctic adventure* and *exceptional Arctic*). It should be underlined that the three case studies in this book discuss a variety of different worldviews and, to some extent, the different interpretations and consequences to which they give rise, which, taken all together, offer a new lens on the science–policy circle in the Arctic.

The analysis in this chapter discusses interview data as well as findings from literature and document reviews. Furthermore, where applicable, we embed our observations in a broader discussion of state-of-the art research. This results in a threefold structure. The focus lies first on how permafrost researchers and glaciologists diagnose geopolitical problems and their causes (Sect. 4.4.1); second, on their perceived prognosis of solution (Sect. 4.4.2); third, and finally, on the values and epistemological foundations underlying their worldviews (Sect. 4.4.3). A summarising table at the end of this section presents the main findings.

4.4.1 Knowing, But With Uncertainties: The Arctic as a Research Object

While the previous chapter analysed the general funding situation of Arctic research in the countries in question, the present one outlines the specific role that permafrost research and glaciology play in Canada, Germany and Norway at both national and international level. This is done by analysing the perspective of the scientists who actually do permafrost and glaciological research. In order to understand the *diagnosis of perceived geopolitical problems and their causes* (see Table 2.1), we approach the Arctic as a research object, by asking what is perceived to be politically and scientifically relevant and why? As the Arctic is a field in which science plays a major role and is visibly prominent, we attempt to unravel the reciprocal relations between science and politics. Here, in particular, we are interested in the researchers' reasons for undertaking their research. We ask, first, about the perceived knowledge gaps in the fields of research they elect to study and reflect on them in relation to published research strategies and agendas. While it is often the same people who are consulted for the development of research strategies and the execution of major projects as Principal Investigators, we are also interested in the perceived knowledge gaps that reach beyond research needs in the natural sciences. Second, regardless of whatever further discussion there might be as to *why* research is relevant and which knowledge gaps exist, this subchapter also focuses on how research is *made* relevant. We provide insights into arguments as to *how* research becomes, or is made, relevant (for scientists, for politics and for society). This also requires asking how scientists understand their work politically, for example, whether there is a political agenda they contribute to, benefit from, ignore or argue against, and how they perceive the larger political and societal framework in which they work. In addition, the question of what role the three countries should or should not play in Arctic research more generally will be discussed.

Bibliometric studies report a steady growth in Arctic research publications since the mid-1990s, dominated by the natural sciences, and embedded in a general global trend of more and faster scientific output (Andreeva & Hønneland, 2023; Aksnes et al., 2023). Arctic research—and thus also permafrost and glaciology projects—arises out of national research strategies, and it turns out that *geography matters* as far as worldviews are concerned: it arguably makes a difference whether a researcher is based at a research institution in an Arctic or a non-Arctic state. At the national level, Canada, Germany and Norway have published and/or updated Arctic strategies in recent years (Canada's Arctic and Northern Policy Framework 2019, supplemented with Canada's Arctic Foreign Policy 2024; Germany's Arctic Policy 2019, updated in 2024; the Norwegian government's Arctic Policy 2021, updated in 2025). In addition, national strategic research agendas (e.g. Polar Knowledge Canada: Strategic Plan 2020–2025; Polar Knowledge Canada: Science and Technology Framework 2020–2025; Geological Survey of Canada: Strategic Plan 2018–2023 and 2023–2028; Mare:N-Begleitkreis: Forschungsagenda 'Polarregionen im Wandel' 2021; German Research Foundation: Polarforschungsagenda 2030;

Research Council of Norway: Norwegian Polar Research, Research Policy 2014–2023) are publicly available. Further, more scientific network organisations as well as Indigenous peoples' organisations have published their research strategies in Canada and Norway (e.g. PermafrostNet Canada: Permafrost Strategy Canada 2023; Canadian Permafrost Association: Strategic Plan 2021–2025; National Inuit Strategy on Research 2018; Sámi Arctic Strategy 2020). At the international level, knowledge gaps and research needs are communicated not only clearly but publicly as well, for instance in the ongoing ICARP IV process (2022–2026), initiated by the International Arctic Science Committee.[2]

4.4.1.1 Knowledge Gaps from a Global Perspective

Overall, the interviews show that research interests and knowledge gaps in glaciology and permafrost research are reflected at different levels: globally, nationally, regionally and locally. Generally speaking, there is no way around looking at contemporary Arctic research in the light of climate change. Starting from a global perspective, not surprisingly, a strong focus on climate change as the overarching concern is apparent. *What happens in the Arctic doesn't stay in the Arctic*—this saying has been repeated countless times, which does not make it any less true. It directly speaks to the worldview of an *endangered Arctic*, a vulnerable region of global importance. Melting glaciers and thawing permafrost can be regarded as powerful ambassadors of this message. Climate change research plays an important role, both locally and globally: on the local/regional scale, for instance, when looking at the threatened livelihoods of Indigenous and local communities, coastal erosion, the collapse of infrastructure built on permafrost, wildfire impacts, water security questions and hydropower from nearby glaciers, glacial landslides or glacial lake outburst floods. On a global scale, permafrost research and glaciology are pivotal for calculating future emission scenarios and estimating sea level rise, especially when calculating the mass loss of the Greenlandic ice sheet. In fact, one glaciologist stated that, regardless of anyone's own narrow research interest, it is not possible to look at glaciers without having climate change in mind, because it has such a big impact {C4_13.4}. In the field of glaciology, the big picture question concerns mass balance measurements of ice sheets and glaciers to estimate their contribution to global sea level rise. In this instance, a regional focus lies on Greenland and Antarctica {C6_4.19}. Nonetheless, the ice masses of much smaller, but faster-melting glaciers in the Canadian Arctic are covered in this context as well {C1_2.10}. Contributing to this research topic, glaciologists investigate ice dynamics, instabilities and ocean–ice interactions, or, in the words of one interviewee, the classical IPCC-topics {G6_9.13}. Closely related, the other major research interest is the use of ice cores as climate archives to reconstruct paleoclimates {G3_5.19 | G6_9.14}, which ultimately leads to a better understanding of anthropogenic global

[2] https://icarp.iasc.info/ [accessed 21 Oct 2025].

warming. In permafrost, the big picture question aims at understanding and estimating how much carbon is converted into greenhouse gases (carbon dioxide), and over what time periods. As the complex thawing and freezing processes in permafrost are not yet completely understood, these feedback processes have not yet been included in the climate model calculations. Therefore, a lot of uncertainty exists about the extent to which climate change will be amplified {G2_7.7 | G1_8.6}.

We find that this corresponds with the research needs outlined in a concept paper on the German research agenda for polar regions. In the chapter 'Ice sheets: Tipping points for sea level' [author's translation], additional emphasis is placed on the societal implications. Given that in 2050 an estimated one billion people will be living near a coast, the estimation of global and regional sea level rise constitutes an urgent research need (Wissenschaftlicher Begleitkreis Mare:N, 2021, 12–15). Likewise, the chapter 'The future of permafrost ecosystems' [author's translation] underlines the importance of estimating future greenhouse gas emissions in a changing climate, understanding carbon feedback processes and their integration in Earth system models, as well as different changes of ecosystem functions (*ibid.*, 20–23). The research agenda is titled 'Polarregionen im Wandel', which translates as 'Changing polar regions', and, accordingly, research needs in the polar regions are described in light of climate change, with a focus on local, regional and global consequences for the benefit of evidence-based policy-making. Unravelling the scientific and political implications of an *endangered Arctic* worldview, we discover a particular self-conception among researchers arguing for German involvement in Arctic research, evident in their describing Germany as a financially strong partner, the 'nice guys', who are solely interested in scientific questions and who are doing a good job in Arctic and Antarctic research {G6_9.5}. Permafrost, as perceived by German researchers, is on the political agenda {G4_6.16}.

Research agendas in Canada are formulated less explicitly and extensively in the specific research fields covered than in either Germany or Norway. Polar Knowledge Canada's Strategic Plan states that its aim is to 'support locally relevant and globally significant knowledge creation' (POLAR, 2020a, 3). POLAR's Science and Technology Framework identifies three gaps: (1) knowledge of dynamic northern terrestrial, freshwater and marine ecosystems needs to improve in the context of rapid change, (2) understanding of the connections between northern community wellness and environmental health needs to be increased, and (3) sustainable energy, technology and infrastructure solutions for the unique environmental, social and cultural conditions in the North need to be advanced (POLAR, 2020b). It seems that the needs of Canadian research in the context of climate change are more targeted towards regional and local questions, as well as more community-driven {C5_20.18}. Framing Arctic research in relation to the climate change narrative appears to carry more advantages for a non-Arctic state like Germany than for an Arctic state like Canada. This takes the analysis one step further by not only identifying the worldview, but also its function. Basically, one could argue that Canada does not necessarily need a climate change framework to advance research on its own territory. Research camps, stations and expeditions in the Canadian High Arctic also prove to be an easy way for the government to establish presence in these remote areas

(Wood-Donnelly, 2019). In addition, the government in power strongly impacts the research agenda {C4_13.8}. However, despite its being an Arctic or 'permafrost country' (Gruber et al., 2023), increasing calls for action towards more coordinated research in Canada can be heard. Supporting the *endangered Arctic worldview*, or maybe even more precisely an *endangered Canada worldview*, Canadian scientific communities actively push for a stronger climate-change-related research agenda and Canadian global leadership on research topics which, in a global comparison, affect the country more than others (Gruber et al., 2023).

For Norway, information is provided by the Research Council of Norway's strategy 'Norwegian Polar Research 2014–2023' (RCN, 2013). In general, research in the polar regions is understood as part of the global system (RCN, 2013, 3), and additional emphasis is given to climate and marine research activities, especially in Svalbard. The brief sections on glaciers and permafrost in the list of thematic priority areas point towards the research questions of future sea level rise and permafrost carbon feedbacks and stability (RCN, 2013, 14–17). All in all, the agenda speaks of research needs in the Arctic or polar regions mainly from a global perspective, within the framework of a 'changing climate and an environment under pressure' (RCN, 2013, 20). References to regions in the Arctic are restricted to Svalbard. Here, the research needs to cover '[p]otential scenarios and critical loads for the development of human activity in Svalbard' (RCN, 2013, 20). It is notable, when comparing German and Canadian strategies with the Research Council of Norway's strategy, that the latter begins its list of priorities with geopolitical issues, followed only later by issues related to climate change.

When did Arctic research become so inextricably linked to climate change? As Mark Serreze recalls in his book *Brave New Arctic*, it was around 1996, in the USA, that the science community started to recognise that changes taking place in the Arctic needed a comprehensive study (Serreze, 2018, 107). A 'Dear Colleague letter' advocating 'A Study of Arctic Change' was put forward by a number of scientists, mainly oceanographers, who were concerned about hydrographic measurements from the Arctic Ocean (Morison et al., 1998). Still accompanied by a lot of uncertainties, for instance natural and seasonal variabilities in sea ice extents, these developments can be understood as the start of an international and interdisciplinary endeavour to link observations in the Arctic to global changes. Back then, permafrost and glaciology topics were not a strong focus of the newly formed Arctic research community. However, Willi Dansgaard had already started to use ice cores for paleoclimate reconstructions through analysis of stable isotopes in 1967. In fact, Dansgaard together with Chester Langway and Hans Oeschger 'led the very first climate research motivated ice core drilling through the Greenland ice sheet at Dye-3 in south Greenland in 1979–1981' (Steffensen, 2012, 206). In 2004, the Arctic Council's Working Group AMAP and CAFF, together with the International Arctic Science Committee (IASC) published the first and still popular Arctic Climate Impact Assessment (ACIA) titled 'Impacts of a warming Arctic', outlining the crucial importance of a rapidly changing Arctic to the world (ACIA, 2005). This constituted a turning point, putting pieces and parcels of different field studies together under the framework of Arctic climate change (Serreze, 2018, 179). Since then at the latest, the scientific urgency of cryosphere topics, and the idea of treating Arctic regional issues in a collective way, has made it into major political discussions. In 2019, the IPCC published its first special report on 'The Ocean and Cryosphere in a Changing Climate' (SROCC) (IPCC, 2019). Since 2009 the International Climate Cryosphere Initiative[3] has been advocating that more attention be paid to cryosphere topics by governments and NGOs. Since COP 25 in Madrid, the Initiative has been organising a cryosphere pavilion at the United Nations Framework Convention on Climate Change (UNFCCC) COPs. However, it took until 2022, at COP 27 in Egypt, for the word 'cryosphere' to be mentioned—briefly—for the first time in the Sharm El-Sheikh implementation plan: '*The Conference of the Parties* [...] *Recognizes* the impact of climate change on the cryosphere and the need for further understanding of these impacts, including of tipping points' (UNFCCC, 2022) {N1_11.11}.

As this book is concerned with *worldviews of ice* and, in particular, ideas of the Arctic, we notice that perceived knowledge gaps are linked to the bigger picture questions of global challenges. From a global perspective, the two major research questions in glaciology—contribution to global sea level rise and ice cores as climate archives—are pivotal in the light of climate change. Global concerns in permafrost research deal with carbon feedback processes and their integration into climate models. In a complex planetary system, the polar regions not only represent sentinels of a warming globe, they also are key elements in preserving crucial oceanic and atmospheric interactions, and avoiding any cascading effects from tipping points. It can be concluded that the cryosphere research community generally pursues similar objectives in terms of what is considered scientifically relevant. At this global level, perceived knowledge gaps arise first and foremost from acknowledging global climate change as the most important issue that needs to be tackled {G5_10.1}. In other words, and following our analytical frame: looking at the Arctic from a global perspective leads to climate change being *diagnosed* as the perceived geopolitical problem.

Research needs embedded in a climate change framework support an *endangered Arctic* worldview. One could now argue that the *endangered Arctic* worldview and the supporting narrative serve as a blueprint—where the importance of the research itself is taken for granted, just, or especially, because it is happening in the Arctic. In '*Frozen archives on the go: ice cores and the temporalization of Earth system science*', Erik Isberg (2022, 279) demonstrates how ice in the form of ice cores, 'in an already mediated form, recovered and quantified', has appeared as a proxy to establish Earth system science. Ice taken from Greenland, for instance, is no longer bound to Arctic geographies, and research is not about the ice itself, but about how it can contribute to an understanding of the Earth system. Following on from these thoughts of ice humanities scholars, one could even go a step further, asking to what extent mass balance measurements of ice sheets to predict global sea level rise are closely entwined with an interest or a concern for the polar regions as

[3] https://iccinet.org/ [accessed 21 Oct 2025].

such at all, and whether they should not rather be seen as embedded in a wider global framework?

First, this could possibly result in an increase in attention and accordingly funding for cryosphere topics and Arctic research. In addition, the *endangered Arctic* worldview speaks to aspects of uncertainty, in insofar as, though a lot of knowledge exists already, many processes are still far from being completely understood. In a world where every scientist has to convince funders of a project's relevance {C2_3.27}, this might appear as a promising way to go in order to sustain one's specific research agendas. Research topics in glaciology and permafrost especially contribute to this idea of an *endangered Arctic*. One needs to keep in mind that science, after all, remains a job for scientists by which they earn their daily bread. They have, therefore, an intrinsic interest in a competitive funding situation in arguing for the importance of their own research elsewhere than only in cryosphere science. Asking about the funding situation in someone's own research field often evokes the response that 'it could always be more', and it is revealing to see how permafrost and glaciology in the Arctic are compared to other research fields, or research in Antarctica {G7_1.20 | G7_1.24}. In the case of Antarctica, O'Reilly observed a behaviour in which Antarctic scientists turned into climate change experts in public presentations. On the one hand, this often fulfils the audience's expectations. Like the melting Arctic, the climate tipping points in Antarctica stand as symbols of climate change, and Antarctic researchers are observing these changes first hand. On the other hand: 'Discovering and reporting on charismatic data about climate change has become one of the most popular justifications for expensive and logistically complicated national Antarctic research programs' (O'Reilly, 2017, 145). This can also be applied to Arctic research. Media attention and public awareness of the key role of Arctic regions enable research funding. As climate change amplifies, it is an easy and successful ploy to frame Arctic research as climate change research, even if this does not do justice to the specific research question. However, evaluations of big flagship programmes like the MOSAiC expedition are more differentiated than the institute's uniform communication strategy would suggest. On the one hand, it is well-received for creating increased attention, interest (and funding) for Arctic issues in general {G4_6.22}. On the other, it strengthens the image of Arctic research being primarily research conducted aboard a big, comfortable research vessel, thus leaving less space and fewer resources for reports from smaller field-based terrestrial expeditions {G3_5.2}.

Second, it can also be questioned, who, given the strength of the *endangered Arctic* worldview and the demand for a strong research framework on climate change, actually still has the opportunity to reflect critically, and decide which research is relevant. If any research in the fields of glaciology and permafrost is uncontroversially considered important, and the researchers themselves are put on the stage as problem-solvers, what are the consequences? While the function of an *endangered Arctic* worldview and research agendas tailored to scientifically responding to the climate crisis is certainly to legitimise research funding, at the same time, critical voices do make themselves heard. For instance, one glaciologist remarked that the loss of biodiversity should be humanity's biggest concern taking priority over climate change {G6_9.16}. Another researcher stated that the decision on research priorities should ultimately lie with the peoples of the Arctic {C1_2.39}. One interesting critical reflection on the importance of people's own research took place in interviews with scientists (permafrost and glaciology) in Canada and Germany. When asked about the relevance of their own research to perceived knowledge gaps, they answered that they could only say anything about their own field of research. Who is to decide what is really worth funding and what the relevant knowledge gaps are? One permafrost researcher admitted that they do not really know what the problems are. All they know is that they need to do something, and therefore they 'follow the calls of publication mills' without really analysing the research needs {C8_12.2}. For natural sciences, the IPCC enjoys high legitimacy in summarising the state-of-the-art science—what is known and what remains to be researched. Some interview partners participated in the writing of IPCC chapters, particularly the Special Report on the Oceans and Cryosphere. But who would in fact be the international authority that could evaluate the necessity of research questions from a global perspective? Though widely cited and highly successful in their respective fields, these interview partners were wondering whether the natural sciences should be put on stage in the way they currently are. Since climate change is considered to be one of the biggest challenges facing humankind, maybe disciplines dealing with society should be put more upfront {G5_10.1}. In a similar context, consideration was also given to the worldviews that generate private funding for research projects. How legitimate are natural science projects funded by philanthropists who are convinced that permafrost is the most important challenge that requires attention, money and publicity? While explicitly not criticising the science behind, one interviewee pointed to the non-democratic nature of such developments {G5_10.3}.

Third, in side discussions at different conferences and workshops, as well as in some of the interviews, a number of glaciologists and permafrost researchers expressed the view that 'pure', original science is still urgently needed. For instance, research on glaciers just for the sake of learning more about them supports the idea of curiosity-driven science {C7_15.8}. The discovery of new things whose practical relevance for society is not immediately apparent, or where the outcome remains open, is, after all, the heart of scientific endeavour. An unbalanced shift towards applied science contradicts this idea, namely that knowledge gaps can be approached regardless of whether they fit into the current socio-political climate. However, while glaciology is a rather young research field, it is possible to observe a change from basic science to more applied or socially relevant research {G6_9.12}, which understands glaciers as a 'climate indicator' {C7_15.8}. In reality, many researchers experience a mix, so that their research is partly driven by climate change, and is partly fundamental research simply intended to find out how glaciers work {C4_13.2–13.3}. Depending on the specific research questions at hand, climate change impacts may remain more or less in the background. Basic research on, for instance, physical processes inside glaciers, how ice streams move or the bio-geochemical composition of organic and non-organic material in the frozen ground, might arise out of scientific curiosity and may be embedded, but may not necessarily be dependent on a climate change research framework. On the other hand, it is simply not possible anymore to ignore the impact of climate change on glaciers {C7_15.7}. In addition, according to one Canadian researcher's assessment, glacier studies with direct links to climate change impacts are more likely to get funded {C7_15.14}. Similarly, researchers' motivation to begin a career in the cryosphere cannot be attributable only to concern about a warming world in which the Arctic constitutes a sentinel. A sense of discovery, childhood memories, the pursuit of a so far little-explored niche, the will to investigate the complexity of scientific phenomena and systems, or a basically pragmatic decision can also lead to a career path as a permafrost researcher or glaciologist.

4.4.1.2 'Arctic' Research at the Regional and Local Level

Closely related to the above discussion about knowledge gaps and the need for basic research in the cryosphere is the question of how researchers deal with uncertainty in their research and what they make of it. As outlined previously, in glaciology one can experience a lot of uncertainty regarding future scenarios of global sea level rise or when global tipping points will be reached. Many physical and mechanical processes inside glaciers as well as microbiological processes in glacial ecosystems are far from well-understood. In permafrost research, huge uncertainties exist with regard to carbon-cycle feedback processes. Nevertheless, the focus on what is yet unknown and what science is striving to discover should not overshadow the generally recognised and well-known global trends.

With a global perspective on the Arctic as a research object, and taking the *endangered Arctic* worldview in particular, one can easily forget about people and the diversity of the Arctic. In fact, one encounters a multitude of smaller-scale, local and regional research questions regarding permafrost and glaciology. Researchers referred to Indigenous knowledge and how their long-term observations of ecosystems, connection and lived experiences with Arctic landscapes lead to a better understanding of local changes {G6_9.51 | N6_18.46 | G2_7.26 | C6_4.37}. Canada, Germany and Norway have defined national research priorities for Arctic research at the global scale. Regional and local research agendas, however, may deviate from this. In fact, the variety and number of small projects, which are not necessarily synchronised, point to there being many open questions that need to be answered at local and regional level. Many reasons contribute to this, such as the complexity of systems with seemingly contradictory trends, specific community-led research needs, or the accessibility or otherwise of places affecting the character of case studies. In addition, researchers are not only working in the Arctic regions. Some started their careers in the Swiss Alps, some regularly travel to Antarctica, others have conducted fieldwork in Iceland, the Himalayas, the French Alps, or the Andes, thus bringing field experiences and diverse research backgrounds with them.

In this world of uncertainty, with complex questions and even more complex solutions, reference to an *endangered Arctic* fulfils an ordering function, as research is structured within the framework of climate change. Related to this, but particularly relevant in argumentation in favour of the importance of one's own research, however, is another perception: that of an *unknown Arctic*. The saying 'the more one knows, the more one knows what is yet unknown' might be appropriate here. Knowing the importance of Arctic regions for the global climate system, and having built up attention and concerted resources for Arctic research, the 'unknown' emerges as something that needs to be understood all the more urgently. Prominent research gaps are formulated in national and international research programmes, or discussed in scientific publications. Interviews as well as conference observations show how many questions remain open and yet to be answered. While the global picture might be crystal clear, making sense of local changes, variabilities and impacts involves a whole different set of challenges {C7_15.11}. One of them concerns the transfer of locally observed phenomena to a regional or even global scale: If it makes a difference to the data when sites are moved by a few hundred metres, then how can you extrapolate across the Arctic regions?

Comparing glaciological studies in Canada, Svalbard and Greenland, for instance, glacier mass loss operates in different time scales. Glaciers in the Canadian High Arctic, that is the Canadian archipelago including Baffin Island, Bylot Island and Ellesmere Island, are much smaller in mass than the bigger ice sheets, and therefore likely to melt earlier {C1_2.10}. While tidewater glaciers in Greenland occur in deep fjords, with warm Atlantic water melting them from underneath, the fjords in Canada are shallow and filled with cold water from the Arctic. Here, the melt rates depend on surface air temperature {C1_2.11}. When asked about priorities in Canada, one interviewee stated that a strong research focus lies on changes in the Rocky Mountains and questions of hydrology and water security, especially for agriculture {C1_2.7}. In addition, local glaciers in Arctic communities serve as drinking water resources {C1_2.53}, and are also part of traditional Inuit trails {C1_2.13}. Glaciers in the surroundings of the St. Elias Mountain range in the Yukon, which represent the largest sub-polar icefield worldwide, host a variety of researchers interested in, for instance, hydrological questions of water drainage, ice stream and meltwater flows, glacial lakes and the general connection to climate systems. The drainage patterns of Kaskawulsh glacier, for instance, abruptly changed due to a climate-change-driven retreat in 2016, which has been documented as 'river piracy'. Before May 2016, glacier drainage divided into one flow to the Kaskawulsh river, a major contributor to Alsek River which drains to the North Pacific, while the other flow headed via Slims River into the Kluane Lake, the largest lake in the Yukon and part of the Yukon River Watershed. Now, with altered flows and the disappearance of the headwaters of the Slims River, the water level in Kluane Lake has fallen, which will eventually turn it into a closed basin with significant implications for surrounding ecosystems (Shugar et al., 2017), as well as for local communities {C6_4.30}.

The Geological Survey of Canada runs the longest continuous long-term glaciological mass balance programme in the Canadian High Arctic, with its earliest series from White Glacier (Nunavut) and Meighen ice cap (Nunavut) starting in 1959/60. One focus for glaciology in Canada's High Arctic therefore lies in maintaining and improving long-term monitoring and data collection {C7_15.4}. The mass balance data is feeding the World Glacier Monitoring Service[4] data base, thus contributing to a huge international data collection effort. In this sense, the Covid-19 pandemic and related travel restrictions were a major issue because researchers lost 2 years of data {C7_15.17}. While the 'Canadian High Arctic has over 200 tidewater terminating glaciers'—only five of them are currently studied {C7_15.5). Therefore, perceptions are that glacier research in Canada is less prominent than in other Arctic research nations {C1_2.8}. Notwithstanding the many research topics left to be discovered, one researcher in Canada shared the impression that it is quite difficult to get the next generation of scientists interested in permafrost research or glaciology {C8_12.28}. In fact, one glaciologist responded to the questions on perceived knowledge gaps in Canada by saying that there are too few people studying glaciers in the Canadian High Arctic {C4_13.1}. In addition to that, researchers from Canada expressed their positive feelings about having a liberal government in power, as it allows for more funding for climate-change-related research {C4_13.8 | C7_15.2}.

While monitoring efforts are challenging in themselves, since they involve assuring at least annual maintenance of instrumentation in regions that are difficult to access, they contribute to continuous knowledge production on how the Arctic is changing. Answering questions about the 'why' seems to be an even more difficult task—one which can be carried out selectively in some places, but cannot be applied broadly. One interviewee described how they have various other interesting projects in mind concerning glacier research on larger timescales with a 'more well-rounded view', but limited resources in both staff and funding prevent them from following up on these research topics {C7_15.18}. To be brutally honest, one could now argue that the work of solving the unknown can in fact be hindered or made more difficult by research within the endangered framework, because funding is easier to get when glacier research is directly linked to climate change. This is a development that researchers have come to recognise over the course of their careers. The interviewee just now referred to recalls how, at the time they started with glaciology, research topics were much more exclusively related to basic glacier research without reference to climate change and how this has obviously changed now {C7_15.6}. Climate change has become the leading issue, simply because 'you can't ignore it anymore'{C7_15.7}. On the other hand, it is also very easy today to relate all glaciological research to climate change in one way or another, in order to make it relevant and turn the spotlight onto glaciers {C7_15.9}.

Permafrost research in Canada mirrors a diversity of research interests. While in coastal communities, such as in Tuktoyaktuk or on Herschel Island, coastal erosion prompts more applied research, field sites in boreal forests and tundra landscapes deal with basic understanding of the hydrologic processes within active layers or, increasingly, wild fire effects {C2_3.42 | C2_3.8}. Related, Indigenous knowledge of permafrost landscapes lacks attention {G1_8.8}. Across Canada, challenges to highway infrastructure or natural hazards such as landslides present an ongoing

[4] https://wgms.ch/ [accessed 21 Oct 2025].

stimulus to research work to do with monitoring, adapting and understanding the thawing of the frozen ground.

In Norway one has to distinguish between research in mainland Norway and research in Svalbard. Research into glaciers on the Norwegian mainland focuses on hydrology in relation to water and energy resources {N3_17.1}, tourism and risk management, all of which could be generally understood as applied rather than pure science. In one glaciologist's opinion, a development can be observed in which the societal relevance and interdisciplinarity of glacier research increasingly constitute prerequisites for its receiving funding {N6_18.11}. While welcomed in principle by a researcher convinced that public-funded research should be directly useful for the public {N3_17.4}, there was no question that a balance between applied and basic research should nevertheless be maintained {N6_18.10}. Research in mainland Norway, however, was not called 'Arctic' research by interviewees, because it takes place below the polar circle {N3_17.3}. In fact, it can be described as smaller in scale and more comparable to high mountain or alpine research, as in Switzerland, especially when it comes to research interests in natural hazard monitoring and prevention. This also means that glaciology and permafrost research in mainland Norway is not necessarily connected to global Earth system science, and is less international in character. Norway's mainland glaciers are monitored by the Norwegian Water Resources and Energy Directorate (NVE), a directorate under the Norwegian Ministry of Energy. Data for surface mass balance and length change are publicly available[5] and, as in Canada, feed into the World Glacier Monitoring Service database. Moreover, researchers and research groups are not limited to work in Norway, but also pursue projects in, for example, Greenland {N3_17.2}.

Svalbard, in contrast, is a climate 'hotspot' in the Arctic {N2_14.13} and attracts widespread international attention as a location for permafrost and glaciology research. Some parts of the Arctic are less unknown than others: Svalbard (especially Ny-Ålesund) is a highly accessible well-covered place, with a lot of Norwegian as well as international research activities going on there. Interestingly, when asked about Norway's geopolitical positioning in Svalbard, researchers assessed Norway's research interest to be strongly politically driven and pointed to the potential for geopolitical conflict and contested sovereignty {N2_14.1}. Despite the fact that this narrative seems to prevail, or perhaps because of it, political scientists point to the myths and misunderstandings surrounding the interpretation of the Svalbard Treaty (Østhagen, 2024).

Germany, as a non-Arctic state, has a different interest in the region. Basic knowledge about mountain permafrost hazards, for instance, is not something that drives research on a large scale. In this sense, regional and local knowledge gaps in permafrost and glaciology are not apparent in national priorities. Rather, local studies are conducted to contribute to the larger picture eventually, or in response to other countries' programmes. In addition, research activities intend to include Indigenous and local research needs and knowledge (Auswärtiges Amt, 2024).

[5] https://glacier.nve.no/Glacier/viewer/CI/en/ [accessed 21 Oct 2025].

Here, the choice of field site might be very much dependent on people's personal connections and how cooperation has evolved historically {G1_8.12}. Some people pointed to the ambivalence of the Alfred-Wegener-Institute as the coordinating authority for German polar research: on the one hand, it is politically desired and consequently supported in terms of infrastructure and funding; on the other, it overshadows a broad spectrum of important research at German universities {N2_14.14}.

4.4.1.3 Dealing with Uncertainties

Generally speaking, science has found its methods for dealing with uncertainties: through recognising and defining uncertainty, through defining thresholds, and through exploring the known unknown. Working to gather more evidence for a theoretical or empirical claim in order to reduce uncertainties in fact 'makes science advance' (Kampourakis & McCain, 2019). How individual scientists deal with uncertainties, however, also depends on their perception of the role of science and the purpose of their research. In terms of evidence-based policy, to maintain trust in science uncertainties should be reduced to the minimum possible, in order to preserve scientific authority, legitimacy and reliability. Scientists argue that good modelling is necessary for good policy choices {N1_11.26}. On the other hand, they also argue that the amount of attention paid to uncertainties in data and figures depends on the research question and ultimately on one's own agenda. This view draws a more pragmatic picture of how science can speak to politics, including the conviction that not so much precision is needed when communicating with policy-makers and conveying a message {N6_18.1}. A frequently cited example of this is the IPCC's role in negotiating the Paris Agreement and the 1.5° or 2 °C global warming threshold set there, as well as the tipping points that 'structure action in both worlds'—science and politics (Beck & Mahony, 2018). From a natural science perspective this was understood as a rather simplified, if not a random, choice of number, but it was sufficient to enable policy-makers to successfully realise a political agenda {N6_18.1}.

In contrast to how global climate politics deals with uncertainties encountered at a global level, the situation appears to be very different with regard to political decisions to be taken at the local level. Monitoring permafrost thaw slumps in inhabited mountainous regions in Norway might be a good example. Agreements on adaptation measures to protect the public against natural hazards need to consider many different factors (political, economic and social constraints) besides scientific measurements, and these predictions influence risk–cost assessments and ultimately the decision-making process {N2_14.5}. While it might make sense from a purely natural science point of view to restrict tourism or to close a road or evacuate villages at an early stage, such restrictions tend to be postponed at a political level, depending on the urgency of the situation.

Whether thinking about global, regional or local research topics, the perception of a yet *unknown Arctic* demonstrates the existence of uncertainties and knowledge gaps which may even go beyond a climate change framework. It strengthens a normative belief that science should persist for the sake of science, thus underlining the importance of basic research. However, in contrast to rather wild nature of polar discovery in the nineteenth or twentieth centuries, the unknown today is brought to light by persistence and longevity, especially in the form of long-term monitoring projects. Often, the unknown speaks to and supports the endangered worldview of the Arctic, and it cannot be emphasised often enough what devastating impacts uncontrolled climate change has and will have on humanity. Sometimes, the unknown stands in conflict with the endangered, because the doing of science is limited in terms of resources, staff and funding. This is apparent in national research priorities too, where little of the 'unknown' is set to be solved other than in a climate-change-, local community- or applied-research-related framework. As a result, the *unknown Arctic* resonates in conversations, but without growing into the strongest argument for more research or more funding. Interestingly, older, more experienced generations of researchers, who began their careers at a time when climate change was not yet playing a central role in research, are more likely to think about the yet unknown {C7_15.19}. Younger scientists relate the still open research questions and ideas almost exclusively to the climate change context, that is the context in which they have started their careers and in which they root their motivations.

This section has diagnosed perceived geopolitical problems and their causes in Arctic research, with a focus on Canada, Germany and Norway. Focusing on the Arctic as a research object and on knowledge gaps in permafrost and glaciology at the global, national, regional and local levels has allowed us to investigate scientists' rationale for Arctic research. Two perceptions stood out: the *endangered Arctic* and the *unknown Arctic*. In elaborating on those two, it was possible to (1) trace the evolution and implications of a climate change research framework for the Arctic in a global perspective and (2) offer a closer look into where and how uncertainties play a role in permafrost research and glaciology on the global, but in particular on regional and local levels. In this context, it can be observed that researchers find themselves in a constant balancing act between applied and basic research. Further, the role of Indigenous knowledge was a point of reference for researchers, with varying degrees of impact on their research practices. The next section will now turn to the prognosis of solution, pointing in particular to science communication, (international) scientific cooperation and data management.

4.4.2 More Communication, More Cooperation, More Data (and Less Politics?): Potentials and Limits of Permafrost Research and Glaciology

This subchapter will elaborate on the *prognosis of solution* (see Table 2.1), how permafrost researchers and glaciologists in Canada, Germany and Norway think about strategies to respond to challenges in Arctic research. While science very much includes politics, it is interesting to observe how and where scientists themselves draw a line between their scientific work and the politics involved. Not surprisingly, natural scientists can be people with rather more or rather less interest in politics, and the degree to which political or societal thoughts are reflected in the daily doing of science varies accordingly. The spectrum appears to be broad, ranging from involvement in advisory boards (mostly by those at later career stages) and very clear political worldviews, to involvement in science activism and the experiences of scientists who do not know much about the current political situation in a country (e.g. the current status of Greenland's independence from Denmark) when conducting field research there.

Reflection on the potentials and limits of both research fields/research practices and possible future developments reveals different worldviews of the Arctic. The following sub-chapter is structured around three aspects that came up during the interviews: communication, cooperation, and data.

4.4.2.1 Science Communication

In the 2021 movie *Don't look up* two astronomers discover a comet that is speeding towards Earth and will inevitably destroy it. Political and economic interests initially prevent their concern from being taken seriously, and their attempts to gain media attention and be listened to for the sake of humanity also fail. As far as climate change in the Arctic is concerned, a lack of political action rather than scientific knowledge can be observed as well. Scientists increasingly communicate with the public or with policy-makers while situating their own research in a larger context. However, they face multiple challenges, and, for one interviewee, *Don't look up* represents a not too unrealistic approximation of the situation today {G2_7.9}. Despite this, or perhaps primarily because of it, science communication is regarded as a key measure to overcome this gap and an integral requirement in funding programmes. If communication is perceived as a solution to challenges in Arctic research, who should communicate what to whom? What is the role of the media? And, ultimately, what is the role of the scientist?

In general, the researchers we interviewed expressed the need to communicate, however with varying degrees of their own involvement. Starting from fundamental role perceptions, one early career researcher was convinced that scientists should speak up and act beyond their narrow research position if they notice a wrong development—even if that might be not a comfortable position to be in {N3_17.5}. This assignment of responsibility hints at how political movements such as Scientists for Future empower themselves (Hagedorn et al., 2019): The perceived responsibility goes beyond basic science communication if required by global challenges such as the climate crisis. It is not strictly necessary, and even hardly possible, to be an expert in all areas. Rather, science in itself is given the authority to be heard. Related to this is a more general wish that scientists' advice should be taken 'a little bit more seriously' by both public and politicians {N3_17.6}. At the same time, this researcher clearly rejected the idea of a technocracy {N3_17.7}. In this context, the

Covid-19 pandemic was put forward as a good example of successful science communication. During those times of crisis, scientists were taken seriously and the public used technical scientific expressions and terms ('flatten the curve') in their daily conversations {N3_17.8}. To their mind, while acknowledging that this was an extreme time of crisis, the (outreach) work of virologists or epidemiologists met with recognition and admiration.

The ideals of democracy often played a prominent role in the interviews. One senior researcher elaborated on the importance of education and research as fundamental pillars of a functioning democracy. Scientists' task, in this sense, is to advance and share their knowledge. Moreover, the interviewee put great emphasis on the responsibility to educate younger generations {N4_19.2}, a perceived responsibility shared by the other interviewees as well. Convinced that society is empowered through science and that, if only people are well enough educated, they will find good solutions.

Regarding challenges to democracy, researchers found that science needs to be better prepared or develop better responses to lobbying. While lobbying as representation of interests is indispensable for a functioning political system and debate, the increasingly unequal distribution of power is a cause for concern. Referring to the oil lobby in Norway, one researcher said that scientists should be 'a little bit more outspoken' in their opinion, even though it requires a 'tricky balance' not to step out of the scientist's role {N3_17.9 | N3_17.42} In another case, the fossil fuel lobby was literally referred to as 'the enemy', and one researcher was convinced that scientific communication as originally conceived is unfortunately not working anymore {G6_9.11}.

In reply to a question as to whether more knowledge is needed or whether existing knowledge needs to be communicated better, responses tended to take the latter view. More communication was perceived to be a good thing in principle, but it needs to be more efficient. However, in times of climate crisis, enabling science communication to navigate between clear messages and personal involvement without making oneself vulnerable remains a challenge. One researcher argued that what counts in the end is that people be convinced by the message {N6_18.2}.

Throughout the discussions, interviewees also reflected on, and reinforced, a fundamental principle and value of science, namely objectivity. Regardless of the mode of science communication in Arctic research, scientists should be trusted to be objective {N3_17.10}. This was regarded as a centrepiece that cannot be jeopardised if the knowledge-based authority that distinguishes science from society or politics is to be maintained. Science activism must therefore be treated with caution. One researcher offered a slightly opposing view. As far as they were concerned, posing normative questions and reflecting on research ethics did not make them a worse natural scientist {N4_19.3}. What can be retrieved from practitioners' self-perception and the positioning of science in society—through modes of communication—might be the precautionary requirement to preserve one's own scientific authority in a world of disinformation and fake news.

What remains an issue hard to solve, however, is that everything that is communicated in reports like the IPCC's, meaning those marked by solid classification, high probability, little uncertainty and what corresponds to the broad consensus or lowest common denominator, is already outdated by the time it is published {G6_9.17}. Some years ago, one interviewee was still wondering about a colleague's project proposal to enhance the quality of political decisions; now that would, in their opinion, make a lot of sense {G7_1.61}. The reference to interdisciplinarity and the question of whether it could eventually bridge the gap between science and politics came up in some of the interviews and, in particular, at conferences we observed.

Turning to the more practical issues of how interviewees in fact communicate their science, a broad spectrum became apparent. Some were involved in writing reports, for example for the IPCC. Some maintain close relations with journalists {G7_1.74} who are then able to deliver more comprehensive reportage about their research. Researchers who work in the Canadian Arctic or sub-Arctic prioritise communication to Northern communities and partners over an audience from the south {C5_20.37}. Many scientists are actively involved in various education formats—through participating in school visits, contributing to educational videos or documentaries for children, through mentoring programmes, public presentations in informal settings—on top of the more usual responding to media requests. Some interviewees had experience of taking film teams on expeditions, while others had been unsuccessful in organising and financing similar accompaniments to smaller fieldtrips {G1_8.14}. Some interviewees feared that they are not as eloquent as others, or younger generations, in dealing with social media {C5_20.2 | G1_8.13}. Early-career scientists were active in education initiatives, for instance with schools to educate students about climate change {C3_21.1}. Increasingly, science projects are teaming up with artists as another outreach method, because arts can be closer to the people. Here, it was important to underline for one researcher that these methods of creative communication should not be used provocatively, but in a rational way, conveying the value of nature, without being too emotional {N6_18.3–18.4}.

Listening to scientists and looking at various institutions' professional science communication departments, it becomes clear that science communication in itself is a growing area of tasks and work. It is no coincidence that being a science communicator requires a special set of skills and training and is a complex job in itself. Although the dissemination of knowledge is acknowledged to be an important part of the work of publicly funded scientists, engagement that goes beyond the average commitment is difficult to integrate into a full working day. In fact, the time spent on science communication cannot be used for publishing and will not yet be honoured academically {C3_21.2 | N1_11.17}. However, actively involving themselves in science communication, whether aimed at policy-makers or the general public, can give scientists additional purpose and help overcome feelings of frustration. While some scientists are doing their science for the sake of science, some are starting to question its purpose, trying to connect their very case-specific detailed numbers to the bigger-picture questions. In this sense, one interviewee reflected that communicating helps mitigate a certain feeling of 'uselessness' {N1_11.18}.

Returning to *worldviews of ice* and ideas concerning the Arctic that are prominent in science communication, the *unknown Arctic* helps to motivate (young) scientists in the research community to get involved in Arctic research. While the *endangered Arctic* certainly serves as driver and motivation for increased effort in science communication, the perception of an *exceptional Arctic* helps by telling exceptional stories. In contrast to the subject area, say, of quantum gravity, permafrost and glaciology are illustrative objects about which good and simplified general science communication should be possible {G7_1.32}. Arctic researchers benefit from having unusual and exciting pictures not to mention 'cool stories' available, which always help in conveying a message.

4.4.2.2 Scientific Cooperation

The narrative of Arctic exceptionalism, its being a region untouched by the geopolitical tensions and conflicts elsewhere in the world, has been seriously called into question not only since Russia's unprovoked war in Ukraine on 24 March 2022 and the suspension of scientific activities in Russia and with Russian scientists. Literature has extensively discussed the three dimensions of science diplomacy (science for diplomacy, science in diplomacy, diplomacy for science) (The Royal Society, 2010; for the Arctic case see Berkman et al., 2017). Generally, scholars testify to a strong international scientific community, with a longstanding history (for instance: International Geophysical Years and International Polar Years 1882–1883, 1932–1933, 1957–1958, 2007–2009) and strong political ambitions under the auspices of the Arctic Council (e.g. the Agreement on Enhancing International Scientific Cooperation in the Arctic, 2017). The majority of interviewees shared experiences of cross-country collaboration with regard to expeditions, article publishing or network organisations.

When asked about the exemplary function of Arctic cooperation for other regions on Earth, one interviewee, however, reacted with surprise, saying that they would have expected Antarctica and the Antarctic Treaty System to be more likely to be mentioned in this context, as the Arctic is 'somewhat [...] free of politics in a way' {C6_4.40}. Circumpolar efforts to coordinate research in the Arctic exist. While science in general can be regarded as an international endeavour, with countless conferences, workshops and co-authored articles as proof, the conditions for international research in the Arctic vary among the three analysed countries.

In Canada, for instance, the Polar Continental Shelf Programme (PCSP), as the major logistics provider in the Canadian High Arctic, only funds Canadian-led teams {C4_13.29}. Research in the Canadian Arctic, for some interviewees, seemed not to be as diverse as it could be {C6_4.41}. Others think it is quite international despite challenging logistics and high costs. In fact, one researcher explained that if you have the money to finance logistical support from the PCSP by yourself, it is easily possible to go wherever you want to conduct research {C7_15.20}. This way, it is not politics but money that determines where research can be conducted and in what manner. In Norway, mainland research is very different to research on Svalbard. While the former can be characterised as much more 'Norwegian', and not 'Arctic' per se, the latter has more of an international character, with the

Norwegian-managed international science village Ny-Ålesund, for example. Arguably, Germany is dependent on international research cooperation and is putting great effort into developing and maintaining international partnerships. While the German-led MOSAiC expedition 2019–2020[6] is an outstanding example of international cooperation, it is not the only one. Apart from big expeditions with high public visibility, such as MOSAiC, and certainly in contrast to marine research, many terrestrial researchers work in smaller groups, often of less than ten people, across the Arctic regions. Given the diversity of projects and questions, the research effort appears to be surprisingly uncoordinated, especially because it covers not only international research activities, but also research activities within one country.

Responding to questions about shared (research) goals in the Arctic research community, the interviewees' responses were more or less unanimous insofar as they referred to the understanding and mitigation of, and adaptation to, climate change. The promotion of long-term, trustful relationships with local and Indigenous communities, as well as with different international partners was perceived to be crucial for conducting research in the Arctic. Admitting to having a naive and to some extent ignorant take on this topic, one researcher stated that, in their opinion, countries' national claims regarding the Arctic should not matter; everyone should be doing research in the Arctic, because developments affect everyone {C6_4.42}. All in all, a push towards or support for more cooperation is seen as a solution. Challenges and crises like the Russian war in Ukraine or the Covid-19 pandemic hinder science. In order to arrive at an understanding of what is happening across the Arctic, international cooperation is needed {G6_9.42}. Further, multiple international logistic networks exist (some of them funded by the EU), to promote and facilitate international research in the Arctic. Scientific network organisations build strong glaciology and permafrost research communities at the national and international level, which are actively used or even managed by many of the researchers we interviewed.[7] Especially as far as small research groups are concerned, network organisations play an important role for exchange with the research community. Besides the normativity of good international cooperation, any research necessarily also requires feasible funding in order to be part of it. For instance, in Greenland or Svalbard, a bias towards wealthy nations, or those who are willing to spend money on Arctic research, is recognisable {G6_9.18}.

---

[6] https://mosaic-expedition.org/ [accessed 21 Oct 2025].

[7] For the permafrost community, the International Permafrost Association (IPA, https://www.permafrost.org/) was founded in 1983, with its secretariat based in Canada. Since as long ago as 1963, the International Conference on Permafrost (ICOP) has been the biggest gathering of scientists in this research field. The Permafrost Young Researchers Network (PYRN, https://pyrn.arcticportal.org/) is an international organisation established within the IPA structure. The Global Terrestrial Network on Permafrost is hosted at the Alfred-Wegener-Institute (https://gtnp.arcticportal.org/). For glaciologists, the International Glaciological Society (IGS, https://www.igsoc.org/) has existed since 1936 as a network organisation. It publishes the *Journal of Glaciology* and the *Annals of Glaciology*. Embedded in the structure of the International Arctic Science Committee (IASC), the IASC Network on Arctic Glaciology holds annual workshops (https://nag.iasc.info/) [all accessed 22 October 2025].

Some limitations were expressed regarding the political scope of networks. While such meetings at the science-politics interface are regarded as very useful, interviewees expressed their frustration that the meetings are merely a process and by no means an action measure. As a result, their own scope for action as scientists remains limited to providing information, and thus, a boundary is clearly drawn between areas of responsibility {G7_1.63}. This, however, might be different in the context of actors who are involved in the transfer as such (e.g. when compiling a summary for policymakers, see Chap. 5).

When discussing the importance and values of an international Arctic research community, Ny-Ålesund was frequently mentioned as an example in the interviews. Imaginaries of Svalbard have recently been investigated, including, but not limited to the Arctic Archipelago as a site of knowledge production (Albert, 2024). One interviewee admitted that they 'maybe naively' believed that Ny-Ålesund is a better place {N4_19.4}. While people come and go, return or leave Svalbard, and situations change, some researchers from Norway even referred to Ny-Ålesund and Svalbard as their second home. One early-career researcher explained how the international group of like-minded people made it easy and fun to continue with research on Svalbard {N1_11.16}. In fact, many early-career scholars were introduced to Arctic research through summer courses at the University Centre of Svalbard (UNIS) in Longyearbyen. Currently 18 institutions are running long-term programmes and research facilities at Kongsfjorden at 79° North, making the former mining town the northernmost year-round high Arctic research station. With four flagship research programmes—atmosphere, glaciology, Kongsfjorden system and terrestrial ecosystems—research in the Ny-Ålesund area is purely natural-science-driven {N6_18.5}. Ny-Ålesund and Svalbard, in fact, can be understood as a 'natural-scale laboratory' {N1_11.3}—or maybe even a small-scale Arctic model. This stands in contrast to permafrost research or glaciology in mainland Norway which is much more similar to alpine research in Switzerland or Austria, for example. Svalbard's being a research hotspot in various senses was cited by one interviewee in Norway as the reason for explicitly not focusing their projects in that area, but rather on the mainland. The accessibility and infrastructure on the one hand and the 'Arcticness' of the research environment on the other make Svalbard 'kind of overcrowded' {N6_18.7}. Interestingly, while Ny-Ålesund is an international site, Longyearbyen was referred to being rather Norwegian, and, with the exception of UNIS (the University Centre in Svalbard), not international.

A possible romanticising of an Arctic place like Ny-Ålesund is something that researchers were well aware of, since this plays a role in reflections on the idea of an *exceptional Arctic*. The feeling of belonging to a certain place as well as to a community of researchers who share experiences reinforces this idea. While in the field, in remote or isolated places, people are dependent on each other, and often build long-lasting personal connections. However, one researcher put up for discussion the argument that research communities in the Sahara Desert, the Amazon or the Galapagos Islands would most probably do the same {N4_19.5}. As long as personal commitment, somewhat harsh environmental conditions, and contributing to something bigger are involved, people will tell their stories. This is not to say that the researchers are not accomplishing exceptional and outstanding work. But thanks to a mixture of historical pride and memories of polar heroes with impressive images of a world that is still only accessible to a small proportion of people, the *exceptional Arctic* continues to be cultivated. To address current debates, it must be underlined, however, that the issues of power inequality and exploitation, discrimination and sexual harassment endemic in all areas of society do not stop at the frontiers of Arctic research. For instance, a growing body of literature reveals reports referring to such experiences in Antarctica (Nash, 2021).

4.4.2.3 Data

The Covid-19 pandemic resulted in 'frozen data' (Duveau, 2021). The permafrost community in particular has suffered from a lack of data with the loss of Russian field sites. This led to the relocation of already funded research projects to the Canadian Arctic, which raised the question of balanced circumpolar measurement and data distribution.

Discussing the implications of the pandemic for research during the interviews, a generation gap became apparent throughout Canada, Germany and Norway. Especially when it comes to networking opportunities at conferences and workshops, or planned fieldtrips, PhD students and early-career researchers were disproportionally affected. For some, the cancellation of expeditions reduced research productivity considerably for a year and a half {C6_4.9}. Looking at the research field of glaciology from a historical perspective, one can see how the discipline has changed since its early beginnings. Back then, glaciology was based entirely on field observations, whereas today some glaciologists carry out outstanding research based solely on remote sensing data {C6_4.10–4.11}. By contrast, in some rather small research areas, such as subsea permafrost, the comparatively small amount of data still leads to more scientific disputes or parties defending their own theories {G5_10.8}.

When conducting the interviews, it became clear that there is much knowledge extant about global trends. It is now 36 years since the first IPCC report was published, and one interviewee stated that even if all polar scientists became out of work tomorrow, decision-makers still had enough information about what needs to be done {G7_1.73}. The question of whether more data is needed at all yielded different responses. One scientist stated that, while the overall story is clear, many processes are far from being understood. For instance, glaciologists still rely on fieldwork because the interesting things are happening inside glaciers, where high-resolution satellite imagery is not helpful {C4_13.21}. By contrast, rapid technological developments also result in 'too much data' {C4_13.18}. Additionally, more and higher-resolution data equals an urgent need for better models to process the data and make sense of the numbers {G7_1.18}. Therefore, data management, data processing, data access and data sharing count as challenges to science cooperation. Another scientist explained that, in theory, they could stop conducting fieldwork and still carry on pursuing a career with the data already available. However, they would not do the science if it were not for the fieldwork aspect {C6_4.13}. To be complete, it should be added that the opposite claim, 'more research needed', especially when uttered from politicians, can certainly be seen as a strategy for policy avoidance.

In practice, researchers are merging different approaches and data with the ultimate goal of creating one overall picture. At least all bigger projects combine field measurements with remote sensing and numerical modelling {N2_14.6}. The perceived relevance of technical development, including the challenge of better, more precise modelling can be found across all three countries analysed. Earth system models aim to represent all the physical, biological and chemical processes of the Earth system, thus reconstructing the past and predicting the future. Interestingly, a disconnect could be observed between the community of fieldwork researchers and that of modellers in how they evaluated the models. Some of the fieldwork researchers shared their thoughts on how very proud Earth system modellers are of their models, even though the models often fail to represent complex feedback processes {G5_10.9}. One permafrost researcher even described themself in a humorous way as the one doing the 'dirty work' when validating data on the ground {G5_10.7}. According to another interviewee current models are 'astonishingly poor' as means of understanding non-linear processes, such as feedback processes {N2_14.7}.

Other glaciologists reported the challenges of fine-tuning the prominent large-scale global climate models to make them applicable to their local glacier {N3_17.11}. Conversely, a comprehensive understanding of processes at one particular glacier, for instance through long-term monitoring in the Canadian Arctic, should, of course, contribute to an—at least regional—upscaling of processes. Ultimately, glacier modelling work, however detailed, aims at improving the variabilities in larger-scale models to enable comparison across glacier regions. But this takes a long time and, given how many field sites and locations exist and how complex and sometimes contradictory the data are, it is hardly imaginable how and when a comprehensive picture of 'the Arctic' will eventually emerge or and whether it ever will at all.

Another problem that was often referred to was the limitations of modelling as a concept. Taking permafrost as an example, one can see that challenges arise with the research field itself. Models can only implement data that is already understood. Permafrost carbon feedback processes are non-linear, thus increasing uncertainties in predictions {N2_14.9}. Permafrost is not as visible as glaciers, as it is under the surface and cannot be easily observed from above. Thawing and freezing take place at microbe level, in contrast to the well-advanced resolution of Earth system models with 100, 50 or even 10-kilometer scales {N2_14.8}. Furthermore, research on very old permafrost cores, for instance, creates unique data sets, but there is no way of telling when these results will find their way into climate or Earth system models {G2_7.11}.

The advance of new technologies, especially in remote sensing, plays an important role, too, when speaking of how to reduce uncertainties and gain more knowledge. Higher resolution in satellite imagery allows for more details to be observed and monitored leading to new questions. As a disadvantage, the better and more advanced high-resolution images become, the more difficult it gets to represent them accurately in the model. Nevertheless, data change science and the development of new technologies are increasingly becoming part of scientific research. For example, the systematic use of drones over otherwise inaccessible parts of a glacier opens up new possibilities of data generation. Technical developments in radar measurement are of particular importance in the Arctic, especially for Norwegian sea-ice and environmental monitoring, as well as for ensuring maritime security {N2_14.10}. Under Arctic conditions, which mean often cloudy weather and half-year darkness, satellites with optical sensors often are of little help. Consequently, the development of cloud-penetrating radar is important for obtaining reliable data throughout the year, for instance in the application areas of ice, oceans and tectonics. As satellite orbits converge above the South and North Poles, receiving stations in the Arctic enable regular and frequent downloading of data, almost in real time {N2_14.11}.

More data ultimately requires more capacity in data processing. While advanced computer capacities and artificial intelligence help in this endeavour, researchers have to deal with more data than they are able to process. They are, with their data and their models, in a way always lagging behind reality.

Therefore, more data is not per se something that helps in solving knowledge gaps in an *unknown* or *endangered Arctic*. Efficient data analysis, data transfer and validation of models are a cross-cutting concern. From an international as well as from a national perspective, researchers refer to a lack of project coordination, data sharing and database maintenance. By contrast, a systematic observation of the Arctic from above, with satellites producing freely available data, in a sense contributes to the democratisation of the process {N2_14.12}. Remote sensing is seen as a measure able to cross borders. In that respect, it fosters the understanding of one Arctic for everyone—at least once people have the technical means and necessary know-how to work with the data. Lastly, in terms of cooperation, one interviewee referred to simple yet equally serious challenges regarding the management of data. In fact, it takes a long time to get from measurement to sorting, analysing and providing the data, especially because, following the logic of science, every research group naturally wants to publish its own data first before having it interpreted by everyone else. Some data is very old and still stored on old hard drives, probably in different formats. These are simple, often institutional hurdles: practical problems, nothing that cannot be solved, but time-consuming and not directly linked to progress. This holds especially true for ice core data (that first has to be drilled for, etc.), compared to remote sensing data. However, there is no evidence of a general unwillingness on the part of researchers to share their data {G3_5.16–5.17}.

Summing up briefly, this section has discussed the potential and limitations of three central elements—framed as solutions—of permafrost research and glaciology in the Arctic: communication, cooperation and data. Beyond that, especially in the Canadian context, community involvement and engagement with Indigenous knowledge were seen as prerequisites for responsible research {C2_3.21}. The *exceptional Arctic* was the third worldview to be prominently featured in this respect. Researchers referred to their perceived scientific responsibilities or the

Arctic as a place of international cooperation amidst geopolitical crisis. In addition, practical considerations can determine the doing of science. Although research often encounters practical challenges that could easily be found in other regional contexts and other research fields too, there is a unique character that permeates the Arctic as a research object. Whether in science communication, scientific cooperation or the role of data, we can observe a circular relationship between science and politics that researchers recognise and—to varying degrees—try to keep their science away from.

4.4.3 From Science to Expertise: Scientists Between Self-Perception and Expectations

This section serves as both discussion and summary of the above, pointing out the underlying *values and epistemological foundations* (see Table 2.1) of worldviews in permafrost research and glaciology: How is the epistemic community reassured about the importance of their research? What is expected from scientists? And what do they expect themselves? In summary, and harking back to our theoretical considerations, can we identify any overarching reasonings that uphold claims for the validity and legitimacy of research?

We turn to concepts of expertise as these help us to understand the roles that scientists impose on themselves and those that are ascribed to them. One important aspect in this regard is that Arctic researchers are working in remote places, in accordance with a worldview based on *Arctic adventure*. Knowledge of the Arctic can, to a certain extent, be accessed through studying textbooks and scientific publications, and a very small percentage of the population actually travels to these places. Arctic researchers in this sense have a special, unique position when conducting fieldwork, because they gain knowledge through 'being there'; knowledge that cannot be learnt but can only be experienced. This argument holds true for both poles, and finds support in Jessica O'Reilly's analysis of the Antarctic as an 'epistemic technocracy':

> In the Antarctic, people gain expertise through living and working on the continent, but that is not the only way Antarctica confers expertise. The mythical properties of the place, the stories of extremes, exploration, and hardship, are carried forward and given to the people who go there. Antarctic people do not just cultivate and sustain expertise; it is conferred upon them through their lives and work (O'Reilly, 2017, 7).

It was remarkable to find out how the importance of fieldwork was emphasised throughout the interviews with researchers in all countries. Those interviewees were strongly convinced that participation in remote fieldwork expeditions not only benefits scientific understanding, but, even more importantly, grows the character, personality and skills of early-career researchers {C6_4.14}. Glaciology and permafrost research, in contrast to, for instance, astrophysics or quantum mechanics, are environmental sciences in which researchers are able to build physical connections with their research object. Observing how researchers on a permafrost expedition in the Yukon got excited about climbing the muddy permafrost exposures to touch ice wedges or to see preserved Arctic ground squirrel nests, enabled one to witness this outdoor enthusiasm that is widely shared among the research community. Arguably, this kind of fieldwork speaks to a notion of *Arctic adventure*, with outdoor clothing, reflective safety vests, safety helmets, bear spray and completed waiver forms for the expedition leader. Adventure here needs to be understood in the sense of authenticity through fieldwork. It is first-hand experience that contributes to the validation of research. Further, a researcher's feeling of love to and connectedness with their field site reminds of holistic Indigenous understandings of landscapes {C2_3.2}. At the same time, in the Arctic tundra landscape of Beringia, it is grizzly or black bears that might be encountered instead of polar bears, it might be 15–20 °C in summer and there may be tourists passing by on the highway. And even if distances feel huge and the lands are vast, people are living and working there, too, with some of them being interested or involved in science and others not {G7_1.5}. This is different to Ny-Ålesund on Svalbard. It today is a designated science village, the population consists of either scientists or logistic providers and, increasingly, hourly tourists. From the infrastructure side, it is an easily accessible place, at least from Europe. It not only has laboratories, but also a canteen, sport facilities, and a shop. Most importantly, this fosters community building activities in a remote place where people are dependent on each other, which makes this place unique and adventurous simply because it is so different from anything else {N4_19.6}. On a field camp in Greenland, by contrast, glaciologists live in tents, after being dropped off by helicopter {G3_5.3}. Listening to the stories of these expeditions, a certain subtle hierarchy can be sensed, and there is a slight subtext of what it means to do 'real' fieldwork. In some cases, however, the decision on whether to camp on-site or to fly to the site from the research station on a daily basis, is simply a matter of funding and the ability to pay for helicopter time. What has certainly changed is the length of time that researchers spend in the field. While some researchers recall times where a glacier was studied for an entire season, this seems—except for overwintering programmes—not to be possible anymore. Contributing to this development are technical advances that enable year-round data collection and monitoring, as well as labour law regulations and the desire for a private life alongside research.

Knowing about a place and understanding that the data requires 'being there'. Becoming an expert might often be a result of acquiring and disseminating specialist knowledge through involvement in committees, through teaching, through projects, in short, through scientific commitment. Being in the right place at the right time, however, helps just as much as developing the ability to have a good, well-received media presence. Across the Arctic prominent researchers can be found who are closely connected to a certain region having contributed to the development of research programmes or logistics and the advancement of science at a certain site. While travelling to the same places and field sites over a long span of time span might hinder a rapid rise in one's scientific career in the beginning, it is also key to unique experiences and the building of strong personal networks and scientific authority. None of the researchers we interviewed could really escape from the fascination and adventure of travelling to Arctic regions, although one admitted that indeed, with ageing, cold and hardship 'hamper the desire to go out in the field' {N4_19.7}. Similarly, family duties or increased desk responsibilities over the course of a career reduce the amount of fieldwork one can do {C8_12.10}. Researchers were not neglecting the positive feelings about a place where their scientific careers started, that one returned to on a yearly basis for a long time, and where memories were created. Nevertheless, the strongly emphasised guiding principle for fieldwork was science, including reassurance about the importance of their work.

Finally, not only in the Arctic science, but in science in general, an extensive debate has begun about how to conduct research in a sustainable, environmentally friendly and responsible way. Bottom-up movements are pushing these discussions forward, critically reflecting on scientists' carbon and environmental footprints. While technological innovations and renewable energies contribute to 'greener' science, their practical implementation throughout remote Arctic regions remains challenging. When one investigates the self-perceived role of scientists in this matter, a generation gap becomes apparent. Despite difficult, maybe even inconvenient discussions about the future of scientific research in the Arctic, senior researchers were very keen to pass on to younger generations the motivation, the joy, the beauty and also, to a certain extent, the light-heartedness of field research. By contrast, despite having a life full of remarkable and exciting expeditions to look forward to, younger researchers tend to be more restrictive, and reflect more on their individual responsibility and travel behaviour (van Soest, 2023).

4.5 Conclusion

What we have presented in this chapter could be described as insights into the black box of Arctic science. Backed by a general understanding of national and institutional structures, a focus on the individual level has allowed an unravelling of worldviews, conditions and practices in permafrost research and glaciology. However, this second case study cannot stand on its own without links to the reconstruction of the Arctic funding landscape (case study 1) and science–politics dynamics as outlined in the Arctic Council (case study 3).

Four *worldviews of ice* have been reconstructed: the *endangered Arctic*, *unknown Arctic*, *exceptional Arctic* and *Arctic adventure*. Worldviews serve to underpin the role-perceptions of scientists, while at the same time illustrating how science and politics are shaped by and actively shape each other—even on the very individual level of how science is done.

This analysis is not, of course, exhaustive—there will be many more worldviews to be found in the research communities. Nonetheless, the empirical foci addressed are meaningful in terms of how scientific research into ice is perceived in the Arctic.

Arguably, the *endangered Arctic* is the most prominent idea, thus influencing the others. Climate change is perceived as a geopolitical issue, and the solution to this

Table 4.1 Worldviews in Chap. 4

Worldviews of Ice	Case study 2: 'Doing Science'
Endangered Arctic	
Diagnosis	Climate change
Solution	More knowledge, including local needs and Indigenous knowledge
Epistemological foundation	Need for evidence-based policy
Unknown Arctic	
Diagnosis	Research gaps
Solution	(Basic) research
Epistemological foundation	Science for the sake of science
Arctic adventure	
Diagnosis	Expectations/motivation
Solution	Media coverage/story-telling
Epistemological foundation	Arctic expertise through 'being there'
Exceptional Arctic	
Diagnosis	Geopolitical crisis
Solution	International scientific cooperation
Epistemological foundation	Science as politically neutral endeavour

diagnosis consists in generating more knowledge. Here, the need for evidence-based policy serves as an epistemological foundation. Interestingly, for the research fields of permafrost and glaciology, and in contrast to the following Chap. 5, *endangered* is understood much more narrowly with reference to climate change. That being said, it also offers opportunities to reflect on local needs and Indigenous knowledge. Looking at the Arctic regions as an *unknown Arctic* gives room for considerations about uncertainties in research. Similar to the climate change framework, a perceived solution results from advances in knowledge and research. In this instance, however, attention is directed more towards basic as opposed to applied research. Here, the basic foundations of science—science as curiosity-driven, science for the sake of science—serve as the epistemological starting point. *Arctic adventure* is informed by a perception that could be generally found across other research communities in other parts of the world. However, it speaks to the expectations and motivations of researchers when thinking about their research from an individual point of view. Field work experience, 'being there', provides scientists with expertise that distinguishes them from others who do not go north. Story-telling is used as a successful tool to promote interest—whether scientific, political, societal or personal. Lastly, the *exceptional Arctic* testifies to a geopolitical crisis in which science is perceived as a politically neutral endeavour, promoting international scientific cooperation and science diplomacy (Table 4.1 provides a summary of these findings).

Chapter 5
Constructing the Arctic III: 'Bringing Science to Politics'

Abstract This chapter turns to the 'translation' of science to politics. It looks at the Arctic Council as a central nodal point in which Arctic scientific and political imaginaries combine, reinforce, or prevail over each other. An important translating process here takes place when results from the Working Groups are discussed and taken up in Senior Arctic Officials and Ministers in Ministerial Meetings. Although deliverables and scientific working plans are agreed, no binding decisions are produced. Therefore, there is no clear overruling power between science and politics worldviews. The chapter briefly contextualise the creation and operation of the Arctic Council from 1990 to 2020. It then focuses on Working Groups in which glaciology and permafrost research projects have been developed, identifying through their reports which Arctic imaginaries and, consequently, which worldviews have been in place. It analyses Senior Arctic Officials' reports and Ministerial Declarations and asks how the identified worldviews from scientific actors have been reflected and how they stand in relation to those from Arctic political actors.

Keywords Arctic Council · Arctic Council Working Groups · Reports · Concept Clouds

In this chapter, we take a look at the Arctic Council as a central forum where science translates to politics, and vice versa. Worldviews that can be reconstructed within science and politics form part and parcel of that translation process as well. However, as should be clear from the preceding chapters of this book, worldviews are not simply 'there', waiting to be discovered or translated between two completely separate worlds of science and politics. Rather, boundaries between science and politics, in the translation process, are constantly made and remade through various representations of the worldviews that underlie or challenge a wide spectrum of relations between these two realms—encompassing cooperation, reinforcement and representation, as well as conflict, fragmentation and suppression. The present chapter analyses this process of making and remaking boundaries between science and politics at the Arctic Council through the different representations of worldviews across

M. Albert et al., *The Worldviews of Ice*, Springer Polar Sciences,
https://doi.org/10.1007/978-3-032-25945-5_5

this spectrum. More narrowly, it focuses on report writing as a practice that configures the boundary which connects these social worlds and establishes an intricate relationship between their worldviews.

5.1 The Arctic Council and Its Working Groups: The Boundaries Between Science and Politics

5.1.1 Antecedents to the Arctic Council: The Arctic Environmental Protection Strategy (AEPS)

With the end of the Cold War and the strengthening of an international environmental agenda, international Arctic governance generally moved from formal to more informal arrangements, which reflected a global trend in many regimes: more transnational and non-state actors were included in order to increase regime efficiency and legitimacy (Tallberg et al., 2013). In 1990, the eight Arctic states founded the International Arctic Science Committee (IASC) as a nongovernmental organisation (Young, 1998; Tennberg, 2000) which aimed to facilitate scientific activity in the Arctic.[1] This favourable cooperative environment encouraged further initiatives such as the Declaration on the Protection of the Arctic Environment (Rovaniemi Declaration) and the Arctic Environmental Protection Strategy (AEPS), both in 1991. Strongly championed by Finland, AEPS envisaged a deepening of the cooperation on Arctic regional environmental issues by promoting general environmental cooperation (Bloom, 1999). In order to facilitate the achievement of AEPS's goals, the Arctic states established four working groups, also including representatives from Indigenous peoples as well as from non-state actors, with IASC being given an accredited observer role (Tennberg, 2000, 80; see Fig. 5.1a, b for two different representations of the timeline of the developments discussed in this section).

Based on the European Monitoring and Evaluation Programme (EMEP), a scientific and policy-led project under the Convention on Long-range Transboundary Air Pollution (CLRTAP), the Arctic states created the Arctic Monitoring and Assessment Programme (AMAP) with a secretariat based in Tromsø, Norway. AMAP (2024) has been responsible for producing reports on the state of the Arctic environment, especially with regard to its pollution, identifying emerging problems and providing recommendations for their management (Young, 1998; Bloom, 1999).

Strongly promoted by Canada in particular, the Conservation of Arctic Flora and Fauna (CAFF) working group is dedicated to the conservation of Arctic biodiversity, supporting the coordination of research and the management of Arctic species and habitat. Based on the elaboration of periodic strategic plans, CAFF has focused on

[1] There is a slight divergence in the description of the governmental nature of IASC, with Young (1998, 34) highlighting its hybrid character due the Regional Boards and their work in assuring that activities were compatible with national interests.

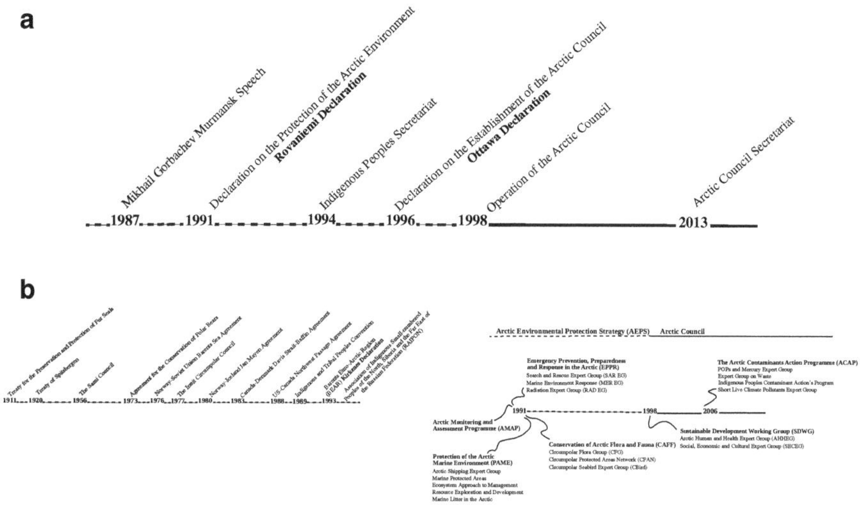

Fig. 5.1 (**a**) Arctic Council timeline—general overview. (**b**) Arctic Council timeline—specific developments

the development of common responses to economic pressures, conservation opportunities and political commitments in relation to Arctic biodiversity. This has also included the development of protected areas and the acknowledgement of traditional ecological knowledge among Indigenous peoples (CAFF, 2025). CAFF's secretariat is based in Akureyri, Iceland, and it initially coordinated three expert groups: the Circumpolar Flora Group (CFG), the Circumpolar Protected Areas Network (CPAN), and the Circumpolar Seabird Expert Group (CBird).[2] In addition, Arctic states' and Indigenous peoples' representatives make up CAFF's management board; observers and non-state representatives join the board during general working group meetings. On these occasions, the group discusses not only its on-going work, but also the development of its long-term workplan and recommendations.

The Emergency Prevention, Preparedness and Response in the Arctic (EPPR) working group has been dealing with the provision of a framework for coordinating responses in case of environmental emergencies, whether natural or man-made (EPPR, 2024). Although EPPR is not an 'operational response organisation', the working group aims to foster information sharing and data collection, developing and implementing activities to support stakeholders on how to be better prepared for emergencies. Therefore, the group deals more directly with subnational levels, especially regarding the need to harmonise national policy responses. In addition,

[2] CPAN was operational from 2006 to 2010. Since then, other projects and programmes have continued the work, such as the Circumpolar Biodiversity Monitoring Program and the Arctic Biodiversity Assessment. CPAN is now dormant (see https://oaarchive.arctic-council.org/collections/2d798074-3b94-4f6b-a37b-852718c4d65c).

the EPPR working group is also responsible for supporting external, legally binding agreements. This includes maintaining the operational guidelines for the implementation of the Agreement on Cooperation on Marine Oil Pollution Preparedness and Response in the Arctic (MOSPA), and direct support for the Agreement on Cooperation on Aeronautical and Maritime Search and Rescue (SAR). With a secretariat also based in Tromsø, EPPR has three expert groups: Search and Rescue (SAR EG); Marine Environmental Response (MER EG); and the Radiation Expert Group (RAD EG).

The fourth entity under the umbrella of the AEPS was the Protection of the Arctic Marine Environment (PAME) working group. PAME was initially dedicated to assessing the adequacy of other international agreements on Arctic marine protection, encouraging the Arctic states to sign and/or ratify any other legal instruments relevant to Arctic marine pollution (Stokke & Hønneland, 2007, 86). Over the years, PAME began to focus on the environmental impact caused by the intensification of shipping activities in the region. Today, PAME defines itself as the centre for 'conservation and sustainable use of the Arctic marine and coastal environment'. Its secretariat is based in Akureyri, Iceland. PAME also has several expert groups covering, such issues as shipping, marine litter, marine protected areas, the ecosystem approach to management, and resource exploration and development (PAME, 2024).

Notwithstanding the efforts to establish effective environmental cooperation, both AEPS and the Rovaniemi Declaration had their limitations: they did not entail legally binding obligations, their scope was restricted to environmental issues, and the substantive inclusion of non-state actors in decision-making was limited. Both aimed to be fundamentally 'non-political', not foreseeing any legal obligations for their signatories. In fact, some working groups, such as PAME, encouraged AEPS members to join other relevant international instruments which actually had legally binding provisions. For instance, although the Rovaniemi Declaration does mention climate change and ozone depletion as 'the two most significant threats' to the Arctic environment, a caveat in the declaration states that these programmes were already being developed by other international fora, whereas AEPS encouraged states to foster institutional links and collaborate on data sharing. Therefore, the legal strings binding Arctic cooperation for environmental protection needed to be provided by these external regimes. Another limiting issue for the governance of the AEPS regime was its wide scope: its backbone was environmental protection and a sustainable environment. No issue linkage was immediately generated when this regime was consolidated. On the contrary: issues affecting cooperation between Arctic states were only prioritised by national policymakers when the latter recognised the impact of Arctic cooperation on their own domestic agendas. Hence, AEPS operations were usually affected by events happening outside its regional cooperative scope.[3] Finally, with regard to inclusiveness, and notwithstanding the

[3]The Barents Euro-Arctic Region (BEAR) was established by the Kirkenes Declaration in 1993, having Norway, the Russian Federation, Sweden and Finland as members. This initiative focused on promoting cooperation on a wide range of issues, such as business, education and environmental protection. Although the Council was made up only of interstate representatives, the Regional

acknowledgement of Indigenous peoples' vital role in the protection of the Arctic environment, AEPS included only the Arctic states as decision-makers. Indigenous peoples' limited participation in the AEPS regime and their reincorporation as Permanent Participants with the establishment of the Arctic Council, illustrate the tension that Arctic regimes experienced when simultaneously widening the circle of participants and the range of the themes that working groups needed to address, but keeping issues restricted to environmental cooperation and decision-making limited to the eight Arctic states.

Although the International Labour Organisation advocated the establishment of the Indigenous and Tribal Peoples Convention in 1989, the organisation of Arctic Indigenous Peoples predates the convention. The Saami Council was founded in 1956 by members in Norway, Sweden, Russia and Finland, aiming to promote their rights and interests within the four countries. The Inuit Circumpolar Council (ICC) was created in 1977 with the aim of representing the interests of Inuit groups in the United States, Canada, Greenland and Russia. In 1990, the 'Association of the Peoples of the North of the USSR' was established to protect their rights and interests, being registered in 1993 as a socio-political movement named the 'Association of Indigenous Small-numbered Peoples of the North, Siberia and the Far East of the Russian Federation' (RAIPON). From 1999, RAIPON became a public organisation uniting 40 different Indigenous peoples. In the first years of AESP, the Saami Council, ICC and RAIPON were only observers in the different working groups. Nevertheless, the necessity of deepening their engagement led AEPS to recommend the creation of the Indigenous Peoples Secretariat (IPS) in 1993. The IPS' creation was the outcome of a shared perception from Indigenous groups as well as Arctic governments (especially Denmark) that indigenous knowledge was part of Arctic environmental cooperation (Tennberg, 2000, 73). In the Rovaniemi Declaration (1991), there were several passages highlighting 'the special relationship of the indigenous peoples and local populations to the Arctic and their unique contribution to the protection of the Arctic Environment', and stating that Arctic states were committed to 'promote cooperation with the Arctic indigenous peoples and to invite their organizations to future meetings as observers' (Arctic Council, 1991, 3). The Declaration also recognises that its environmental protection strategy was prepared with the help of Indigenous peoples, based on initiatives already taken nationally and by the Indigenous peoples themselves (Arctic Council, 1991, 6). In the 1993 meeting in Nuuk, the Icelandic government proposed an Indigenous knowledge seminar and the Danish government agreed to financially help the establishment of a secretariat to facilitate cooperation between Indigenous peoples' organisations (Tennberg, 2000, 73).

Council included subnational bodies and Indigenous peoples (Young, 1998). BEAR was an example of broad-issue Arctic cooperation, albeit in a more restricted geographical area when compared to the Arctic Council. After 20 years of the Council's operation, following its invasion of Ukraine and subsequent suspension, the Russian Federation decided to withdraw from the Council in 2023, which raises questions as to the future of BEAR's effectiveness in promoting cooperation and problem-solving in the Barents region.

Concomitantly to the discussions on how to better engage Indigenous peoples, the Arctic states also perceived the necessity of widening environmental protection by incorporating 'sustainable development' as one of the main Arctic concerns (Young, 1998, 41; Scrivener, 1998; Bloom, 1999). This does not mean that this incorporation was done without initial resistance, mainly from the US due to fear of spilling over into other issue areas (Young, 1998, 63; Tennberg, 2000, 2). In the Nuuk Declaration of 1993, Arctic states affirmed that 'environmental protection constitutes an integral part of the development process and cannot be considered in isolation from it' (Tennberg, 2000, 94). Nevertheless, the definition was not quite precise and an underlying duality between environmental protection and sustainable development endured and delayed further institutionalisation of AEPS (Tennberg, 2000, 110).

Although AEPS was acknowledged as a Finnish initiative, Canada became enthusiastic about this initiative during its negotiation in 1989–1991, whereas the United States showed more resistance against full engagement (Young, 1998, 106).[4] Even during AEPS's first years, Canada continued to advocate for the creation of an intergovernmental forum where both environmental and economic development issues could be collectively discussed by Arctic states. Pressure also came from non-governmental organisations. During discussions on a future Arctic Council, the World Wide Fund for Nature (WWF) emphasised the integral relationship between environmental protection and sustainable development, possibly jeopardised by the Council operating under a false antagonism between cooperation for the use of the Arctic, on the one hand, and its protection and sustainable use, on the other hand (Tennberg, 2000, 110). As a result of a number of factors, including a response to requests for scope-widening, a change of policy in the United States,[5] as well as a more structured representation of Indigenous peoples through the IPS (as sustainable development was intrinsic to Indigenous peoples' concerns), AEPS established a Task Force on Sustainable Development and Utilisation (TFSDU) in 1994, until it was elevated to the status of a working group (WGSDU) in 1996 Young, 1998, 40). The WGSDU was supposed to continue its mandate until the future Arctic Council could establish its own sustainable development programme (Scrivener, 1998, 53). The main goal was to terminate the 'two-pillar imaginary', so that sustainable development would not undermine environmental protection (1998, 54).

This momentum led Canada to propose the incorporation of AEPS into a new international body in 1995. This expansion movement resulted in the Declaration on the Establishment of the Arctic Council (Ottawa Declaration) in 1996, which created the Arctic Council. The new intergovernmental forum co-existed with AEPS for another 2 years, as it progressively absorbed the existent institutional

[4] Although Canada advocated for an international organisation, the United States struggled with such a formal entity, eventually agreeing to the creation of a council without legal personality (Bloom, 1999, 714).

[5] In 1994, the United States published their new Arctic Policy, which not only provided the basis for their engagement in Arctic international forums, but also fostered institutions of cooperation (Tennberg, 2000, 2).

framework, holding the Alta Ministerial Meeting as the last one in the framework of AEPS 1997 (Rottem, 2020, 4). As Indigenous Peoples were able to lobby beyond their nation-state, exerting internal and external pressure through other external bodies (Coote, 2016), the Council, in the Ottawa Declaration (paragraph 2), defined the three Indigenous peoples' organisations (Saami Council, ICC and RAIPON)—who were already Permanent Observers at AEPS—as Permanent Participants and acknowledged that IPS would continue under its framework thereafter. The Declaration also contained a provision for other Arctic organisations of Indigenous peoples (with respective majorities of Arctic constituencies) to join the Council.[6] The AEPS was finally subsumed by the Council in 1998, incorporating the four original working groups and establishing two new ones over the following years.

The Sustainable Development Working Group (SDWG) officially became operative in 1998 alongside the Arctic Council. Its main goal is to foster initiatives that 'enhance the environment and the economies, cultures and health of Indigenous peoples and Arctic communities', focusing at the same time on the practical knowledge of Arctic communities that promotes their capacity to respond to the challenges of the Arctic region (SDWG, 2024). The SDWG has two expert groups: the Arctic Human Health Expert Group (AHHEG), and the Social, Economic and Cultural Expert Group (SECEG). The sixth working group was founded in 2006, a few years after the establishment of the Arctic Council. The Arctic Contaminants Action Program (ACAP) originated from the Arctic Contaminants Action Plan, which was adopted in 2001 and focused on persistent organic pollutants (POPs) and mercury (ACAP, 2025). Today, ACAP is one of several expert groups which are focused on different pollutants such as the POPs and Mercury Expert Group, the Expert Group on Waste, the Indigenous Peoples' Contaminant Action Program and the Short-Lived Climate Pollutants Expert Group. Like all the previous working groups, SDWG and ACAP operate within the Council's framework and provide periodic reports about their activities to Senior Arctic Officials (SAOs).

5.1.2 *The Arctic Council*

The Arctic Council was created as an intergovernmental forum for promoting cooperation and coordination both among Arctic states and between these states and Arctic Indigenous peoples' organisations on issues of common interest. In addition to its core membership, the Council has admitted a growing number of observers, including non-Arctic states, intergovernmental organisations, and NGOs. Focusing

---

[6] Other Indigenous organisations with a majority of Arctic Indigenous constituencies could join as long as they represented a single Indigenous people resident in more than one Arctic State; or more than one Arctic Indigenous people resident in a single Arctic state. Therefore, Aleut International Association, Gwich'in Council International, and the Arctic Athabaskan Council became later permanent members of the Arctic Council and part of the Indigenous Peoples' Secretariat (https://www.arcticpeoples.com/#about-ips).

initially on environmental protection and sustainable development, the Arctic Council's scope has enlarged, addressing other social and economic issues. Notably, however, the mandate of the Arctic Council excludes hard security issues (Rottem et al., 2020, 288). Although the symbiosis between science and politics makes the distinction difficult, the Arctic Council has created specific forums where science takes the fore, and others in which politics is centre stage. Ministerial meetings are the highest level of Arctic decision-making: Arctic states are represented by their foreign ministers, addressing Arctic issues through consensus among member states and in consultation with permanent participants (i.e., Indigenous peoples' organisations). At this level, there is no strategic planning, and priorities are defined by the chairmanship, which changes every 2 years. As the chairmanship rotates among Arctic states, their priorities tend to represent their national aspirations for the Arctic (Bloom, 1999; Rottem et al., 2020).

If at one end of the spectrum the ministerial meetings are highly political, working groups are at the other end with science taking the fore. Working groups have their own secretariat and their own strategic programme, establishing their specific goals which may or may not follow the priorities of the different chairmanships. They are usually represented by environmental ministries and institutes from Arctic states, but participation varies according to their different interests in specific themes. Arctic states and observers provide the scientific personnel and resources required to develop working group projects on a voluntary basis. They therefore only engage with working groups that address topics in which they have a direct interest. While participation in ministerial meetings is restricted to Arctic states and permanent participants, observers in working groups can engage more directly by proposing projects in association with Arctic states or permanent participants, which entails a more active participation on scientific themes (Barry et al., 2020).[7]

At the intermediate level between working groups and ministerial meetings, or, between an extremely scientific and a substantive political forum, are the Senior Arctic Officials (SAOs). With representatives from the foreign ministries from Arctic states and permanent participants, SAOs act on behalf of ministers and guide the work of the Arctic Council's subsidiary bodies—working groups, expert groups and task forces. SAOs form a bridge between the political level and the scientific one by providing Foreign Ministers with progress reports from subsidiary bodies on deliverables achieved in the previous 2 years and their workplans for the next two. SAOs' reports are considered by ministers and confirmed through ministerial declarations, which are non-legally-binding agreements endorsing the deliverables in question,

[7] By 'participation' we here mean an engaged involvement, where actors have power to define the outcome of an event. One could say that observers 'participate' in the Arctic Council, because they can be invited to attend the meetings, give statements after Arctic States and Permanent Participants, present written statements, and even submit relevant documents providing their views on some issue (Arctic Council Rules of Procedure 38; Arctic Council, 2013). However, the continuation of their invitation remains subject to review by. and consensual agreement of, Ministers every 4 years, and they cannot make decisions. Although Permanent Participants do not have decision-making power, their presence in every forum is—like the name suggest—permanent.

while also approving working groups' future working plans. Declarations also highlight emerging issues and instruct SAOs on their subsequent actions. By collecting the outcomes of working groups and guiding their work through the instructions in ministerial declarations, SAOs articulate Arctic worldviews from different scientific projects and inform the Arctic Council's perspectives on what their subsequent actions should be.

However, one cannot easily identify a causal relation between the main takes from ministerial declarations by the Arctic Council, on the one hand, and Arctic national research and funding agencies, on the other hand, especially when it comes to the extent to which ministerial declarations would inform research priorities. The hortatory character of the Council's declarations and SAOs' reports does not establish a direct link between them and national science policies. Nevertheless, they provide a point of reference for framing legally binding agreements in other fora, which means the Arctic Council functions as a 'policy-shaping' rather than a policy-making body (Rottem, 2020). Moreover, the different worldviews about the Arctic underlying declarations by the Arctic Council could be adopted by funding and research agencies or find their way into the methods and approaches adopted by funded scientists (although no direct evidence has been found so far). Issues prioritised in ministerial declarations as urgent and in need of more investigation could influence the research priorities of national funding and research agencies, thus determining the funding of Arctic research projects. On the other hand, ministerial declarations endorse SAOs reports which present the outcomes of projects from working groups. Therefore, Arctic states and observers are most likely to fund research projects that contribute to working group projects with which they are already involved (Barry et al., 2020).

With its three-level governing forums, the Arctic Council, exemplifies the dynamic relationship between science and politics. As a boundary organisation (Spence, 2017), it provides a space where different Arctic worldviews meet, and where a systematic process of translation between science and politics takes place. Reports from working groups to SAOs, and from SAOs to ministers are processed in a procedure during which science is translated to a policy 'lexicon'. The communication of the scientific findings carried out by those on the ground needs to be intelligible to those discussing the Arctic in a broader perspective. For instance, the Arctic Climate Impact Assessment (ACIA, 2005) acknowledged the potential confusion that assessments of the likelihood of events and conditions cause, as they are based on a range of probabilities and reflect evidence from 'field and laboratory experiments, observed trends, theoretical analyses, and model simulations' (ACIA, 2005, 7). Therefore, based on US National Assessment Team and the IPCC reports, the authors elaborated a five-tier lexicon of terms that are 'consistent with everyday usage' and simplified the message they intended to convey (see Fig. 5.2).

If translation from science to politics takes place through the different groups within the Arctic Council, this has consequences for the extent to which Arctic worldviews shared by scientists either fade away and get 'lost in translation' through this reporting process or, in contrast, are reinforced in terms of authority and legitimacy. In other words: is an *endangered Arctic* worldview, as outlined in a working

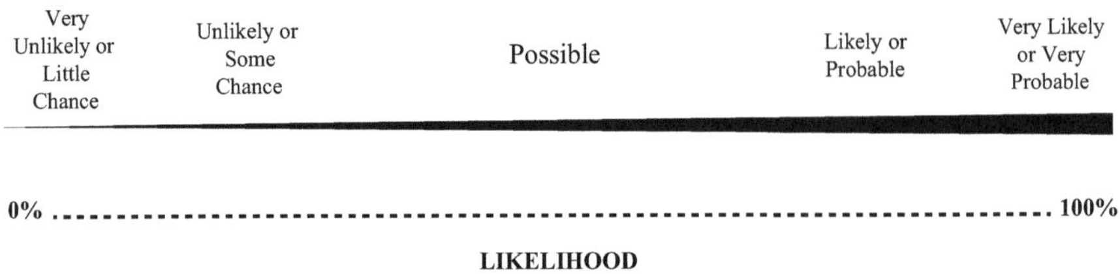

Fig. 5.2 ACIA Five-tier lexicon describing likelihood of expected change. (Created by authors based on the figure in ACIA, 2005, 7)

group report, still fully depicted in a ministerial declaration? Or does an 'Arctic in the Anthropocene' worldview take the fore once we move away from science and closer to a political realm, where an analytical view gives way to a more ambiguous one, and where a negative connotation can co-exist with potential positive human-nature interactions?

This chapter focuses on understanding the extent to which working group reports operate as boundary objects and their authors as 'passage points', linking Arctic science and politics through the translation of their different worldviews. By observing the movement of different worldviews through the Arctic science–politics spectrum, we identify which ones prevail in specific working groups; and how some vanish in the translation process. In the next section, we present our analytical framework, defining the concepts of 'boundary objects' and 'obligatory passage points'. In the methods section, we explain our methodological choices, providing the criteria for the reports' coding and analysis. Based initially on interviews with SAOs, report editors, and working group secretariats and chairs, we mapped the Arctic Council reporting system, identifying the obligatory passage points. We then compiled ministerial declarations, SAO reports to ministers and working group progress reports to SAOs from 1996 to 2021, as these represent the boundary objects at each obligatory passage point. Using the atlas.ti software, we ran a 'concept analysis' to identify the main themes at each passage point, constructing the worldviews that are most representative of each group of actors (ministers, SAOs and working groups).

In the analysis section, we observe how these worldviews were maintained and carried over from one report to the other. We begin our analysis with the ministerial declarations, because our focus is not on what was 'lost in translation' from working groups, but on what was actually transformed and changed through translation processes. A passage point is better illustrated by worldviews that survived, even in a modified version, rather than by those that did not endure the science–politics trajectory. In our analysis, we identified *co-operative Arctic*, *national Arctic*, *endangered Arctic*, *human Arctic*, and *technical Arctic* worldviews. We also explored differences in the worldviews shared by working groups, as some are more prominent than others, contributing to the formation of specific group identities. For instance, 'the Arctic as Indigenous homeland' is likely to be far more prominent within the Sustainable Development Working Group (SDWG) than within the

Arctic Monitoring and Assessment Programme (AMAP), which makes a far more technical contribution, as it focuses on problems whose causes are located way beyond the Arctic. In the final section, we make some concluding remarks on how worldviews were transformed through their translation processes as they crossed science–politics boundaries.

5.2 Bringing Arctic Science to Arctic Politics

From the inception of the Arctic Council, its member states established working groups to provide expert information about Arctic environmental issues (Haavisto, 2001). In fact, working groups have been the forum 'where the majority of the Council's work is undertaken' (Barry et al., 2020, 3). Working groups form an exclusive scientific space, where experts examine and report on specific research topics according to the mandate the group receives. At the other end of the spectrum, ministerial meetings represent the highest political level, with representatives of Arctic states agreeing on the foundations of international agreements and defining the priorities that are of common concern in the region. Therefore, the Arctic Council would seem to be divided—at first glance—into a scientific 'bottom' level and a political 'top' level. However, the seemingly clear boundaries between Arctic science and politics as expressed in these levels become blurred when we consider the role of SAOs: they operate as intermediaries between the two levels, reporting to ministers on the progress of working groups, while also guiding the latter's work in accordance with the priorities of ministerial declarations. SAOs thus act as 'passage points' on a mesoscale: they play a pivotal role in demarcating the boundaries between Arctic science and politics.

Given that the scientific work comprises heterogeneous groups of scientists and non-scientists, we follow Star and Griesemer (1989) who propose an ecological analysis where cooperation among different viewpoints is necessary, so as to connect different social worlds. For these authors, 'scientists and other actors contributing to science translate, negotiate, debate, triangulate and simplify in order to work together' (Star & Griesemer, 1989, 389). Their model expands on Callon and Law (1982) and Latour (1987), who argue that specific actors gather different concerns from a wide range of other actors and re-interpret them according to their own goals. By acting as gatekeepers, they build their own authority, making themselves an *obligatory* passage point. Nevertheless, some level of integrity for audiences' concerns has to be maintained in the reinterpretation: if audiences perceive themselves as disconnected from gatekeepers, if their concerns are no longer recognised, they could find themselves disenfranchised, leading to a loss of gatekeepers' authority and privileged position as passage points.

Although Star and Griesemer's ecological approach acknowledges multiple sets of translation and negotiated passage points, as in an actor-network framework, Callon and Law and Latour narrow down their analysis to the *obligatory passage points*. Within the Arctic Council, several negotiating stances are adopted: within working groups, by SAOs, and by ministers. Working groups negotiate the final versions of their reports and summaries for policymakers; SAOs and permanent participants negotiate the summary reports from their annual meetings and their report to ministers; and ministers negotiate their declarations in their biennial meetings. In this context, Star and Griesemer's actor-network model, combined with Callon and Law and Latour's concept of obligatory passage points, provides a helpful analytical framework, as reporting in the Arctic Council takes place at multiple levels and follows an obligatory route (see Fig. 5.3).

On the other hand, if we look closer at these different passage points, focusing on their scientific components, we can observe that the translation from science to politics is a continuum. But how do actors form this spectrum? In the passage from assessment reports to progress reports, science and politics are intertwined, mostly unnoticed by authors, blurring the boundaries that actors try to keep intact. Nevertheless, within such passage points, differentiating science from and associating it with politics take place through a translation process where logics of equivalence and difference are constructed and whose tension cannot be reconciled. Within the Arctic Council, the differences between working groups (science), SAOs (passage points), and ministers (politics) are suppressed but never extinguished in the recommendations they pass on to each other. This rather fuzzy picture is confirmed not only in the literature, but also in interviews with working groups' secretariats. For instance, not only scientists belong to epistemic communities that will share a particular framing of the questions they attempt to answer, but as representatives of their national delegations they also present some degree of closeness to political decision-making, with their work also rooted within particular national science policies (Tennberg, 2000, 69). This diverse scientific environment was confirmed in conversation with report editors from working groups in 2024, where the multiple reviewing process of reports appears as a 'technical' issue also interspersed with power dynamics. It is not possible to have an Arctic science subsumed into politics,

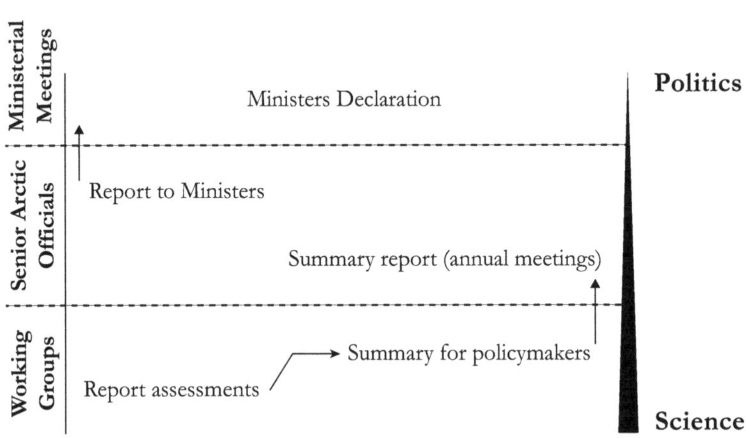

Fig. 5.3 A representation of the reporting dynamics within the Arctic Council

because its legitimacy stems from its perceived 'apolitical' character. Likewise, it is not possible either to represent science's findings literally in communications with political audiences, because their technical character demands a high level of expertise for its comprehension. Consequently, worldviews shared by science actors are not necessarily received when communicated to political ones. Although internally preserved, the boundaries between the different levels are externally blurred by the necessity of translation. If *boundary objects* represent 'means of translation by inhabiting intersecting worlds' (Fujimura, 1992), in the Arctic Council the 'reports' constitute the boundary object between these two Arctic social worlds and the multiple worldviews that they entail.

Boundary objects were originally defined as 'objects which are both plastic enough to adapt to local needs and constraints of the several parties employing them, yet robust enough to maintain a common identity across sites' (Star & Griesemer, 1989, 393). Considering the science–politics continuum, reports from working groups to SAOs, and SAOs' reports to ministers perform a connecting role, in which they change according to the different levels across which they are transmitted, but, at the same time, keep a recognisable structure for those that originally drafted them. However, there are two important caveats to this process. First, working groups vary according to the degree of scientific work they undertake. Whereas some are deeply involved in assessment and monitoring tasks; others are far more policy-oriented (Rottem, 2020), creating different science–politics dynamics. Second, every translation process entails a loss. Translation involves not only an 'irreconcilable tension' (Machen, 2018), which implies an intrinsic loss of knowledge once it takes place, but when information is summarised, worldviews that were once dominant within a working group can also get sidelined once reporting reaches SAO and ministerial levels.

Within the Arctic Council's various reporting practices, it can be quite straightforwardly affirmed that the detailed and technical lexicon is discarded as soon as reporting needs to move towards a non-expert audience. Working groups represent the most technical level at which science is progressively translated through its reporting: scientific findings must be made intelligible to political forums. Highly technical assessments are thus summarised through an iterative reviewing process involving editors, working group secretariats and leading authors, with the latter's expectation being that their worldviews will be preserved once transmitted to the SAOs in an accessible fashion. And the connecting element that crosses these different social worlds, navigating their different worldviews, are the reports. As boundary objects which traverse obligatory passage points, reports are the blueprint where some worldviews endure, some vanish and some are transformed.

As passage points reinterpret specific concerns as they pass from one group to the other, the immediate question that arises is the extent in which worldviews are modified on the way. If worldviews are 'narratives about what is possible, what is worth doing, and what needs to be done, as well as what is impossible, what is shameful, and what needs to be avoided' (Katzenstein, 2022, 9), we should expect them to disappear, endure and/or be transformed as they pass through obligatory passage points. And if we unfold this argument, we should also ask if transformation occurs due to the simple necessity of connecting different social worlds in an intelligible way; or if there is also a power dynamic involved which privileges one worldview over another. For instance, when drafting reports, scientists need to negotiate and select what they believe is the most important information for policymakers' consideration. Therefore, politics already permeates this scientific realm, as the inclusion/exclusion of information is an interpretative process informed by the different worldviews of leading authors. By synthetising the information to be communicated, working groups and SAOs' reports converge to a specific worldview not only of the Arctic, but also of the science–politics dynamics, forging a specific identity for each group.

The construction of a report becomes the point at which different Arctic worldviews are represented through an iterative translation between science and politics. Each working group may establish its own operating guidelines, but as they all report their progress to the SAOs, how working group worldviews come to be represented in SAOs' reports to ministers configures a process of science–politics translation. Likewise, ministerial declarations not only provide a high-level endorsement for any information provided from SAOs reports that they encompass, they also mark the direction in which working groups should move. In the next section, therefore, we detail how we analysed the different reports, identifying their respective worldviews and how they configure the boundary objects within a science–politics translation.

5.3 Methodology

As our aim is to understand the translation process between Arctic science and politics, our analytical framework is based on three pillars: the content to be translated, the work of the translators and the outcome of the translation itself. Since Arctic Council reports constitute boundary objects, the translation of assessment reports into summaries, which then form the SAO reports to Ministers, offers a unique insight into the relationship between these two social worlds and especially how Arctic worldviews are retained, transformed and/or erased. Although there exist several passage points throughout the Arctic Council reporting system, the two most influential obligatory ones are positioned within working groups and between working groups and SAOs. Ministerial declarations, as a rule, tend to corroborate rather than modify what SAOs' reports already bring to the table (Rottem, 2020).

The Arctic Council Secretariat makes all ministerial declarations, SAO reports and working groups' assessments available on their website. Hence, we are able to analyse and compare the differences between the ways these actors have communicated information to each other throughout the years the Council has been in operation. Nevertheless, as we move from the most political to the most scientific level, the volume of reports increases exponentially. Since our goal is to observe how translation takes place through observing worldviews being translated from one social world to the other, we needed first to define how to approach the reports. The reports that are situated in obligatory passage points are: working group progress reports—as they need to translate their scientific findings for the benefit of other political audiences; SAO reports to ministers, which provide the foundations for ministerial declarations; and ministerial declarations per se, which guide the work of working groups in the following years.

We initially undertook informal talks with personnel from both the Arctic Council and the working group secretariats in 2022 and 2024 respectively. We also talked to a Senior Arctic Official in 2024. These talks were fundamental to acquiring information about drafting, editing and reviewing practices that are usually not displayed to the public. In a nutshell, drafts prepared by the working group secretariat circulate among the different scientific, Indigenous and national representatives, until a final version is tacitly accepted by members. The report is then summarised with the support of the secretariats, drawing up progress reports and a 'summary for policymakers' that will then be taken on by SAOs. Although working groups have slightly different processes for the elaboration of their reports (AMAP and CAFF are those with the most elaborate ones), these reporting practices are enlightening with regard to how different worldviews can prevail over others from the working group level, because not all contributions are taken into the final version of a scientific progress report. Once we had an informed understanding of reporting practices, we were then able to start a thematic analysis of the different reports.

Based on qualitative content analysis (Schreier, 2012), we created a coding frame to analyse the different reports from the Arctic Council and identified the representations of worldviews characterising the boundaries between Arctic science and politics. We started our analysis with ministerial declarations from 1996 to 2021, as our goal was to identify translated worldviews (not those that never managed to be represented at the highest levels). This approach was also helpful in practical terms, as ministerial declarations comprised the smallest volume of reports, thus allowing a more nuanced analysis. We ran *word frequency*, *concept* and *opinion mining* analyses. The word frequency analysis provided a general impression of the main reiterated ideas within the declarations. The concept analysis allowed for a more substantive mapping of those ideas, particularly through also establishing associations. Word clouds of the words surrounding the different concepts helped us to select those that could be grouped together (as well as those that could be discarded since they did not make any analytical contribution; see Fig. 5.4).

The opinion mining function helped us to identify synonyms and associated concepts. The most represented concepts led us to shape a specific Arctic worldview within this governing level, exemplifying it by declaration passages which comprise the essential meanings of these worldviews (see Table 5.1). We repeated the same process for all Senior Arctic Officials' reports from 1998 to 2021 and for each working group progress report and workplans from 1998 to 2021. Considering that reports constitute boundary objects, the translation of information from progress reports to SAOs' reports, and from the latter to ministerial declarations

Concepts	Count
Work	194
Cooperation	145
Development	141
Change	130
Climate	125
Meeting	121
Importance	101
Effort	97
Community	95
Report	93
Activity	90
Project	87
Need	85
Action	82
Appreciation	82

Fig. 5.4 Ministerial Declarations concept cloud from 1996 to 2021

Table 5.1 Worldviews and related concept coding for ministerial declarations 1996–2021

Worldviews (categories)	Coding references
Co-operative Arctic	Cooperation; collaboration; dialogue; contribution; coordination
Human Arctic	Community; traditional; indigenous; local; development; peoples
National Arctic	Arctic states; government; country; state
Endangered Arctic	Challenge; issue; risk; threat; danger; vulnerable
Technical Arctic	Work; activity; effects; impact; expert; knowledge; assessment

provides a unique opportunity to observe the relationship between these two social worlds.

In the following, we provide a number of quotes that document the respective worldviews in an exemplary fashion:

> *Cooperative* Arctic: '[The Arctic Council] emphasizing the importance of circumpolar and international cooperation as being fundamental to addressing circumpolar challenges' (Arctic Council, 2006, 2).

> *Human* Arctic: 'The meeting reaffirmed the role of the Arctic Council in promoting sustainable social and cultural, economic and environmental development in the Arctic, and in harnessing the economic potential of the region for the benefit of Arctic inhabitants and communities, and looked forward to the opportunities provided by wider application of innovative, resource efficient circular economy in the Arctic' (Arctic Council, 2019, 16).

> *National* Arctic: the Arctic States, in view of the intensified need for global and regional action, will continue to collaborate closely in international fora on environmental protection

and sustainable development issues of importance to the Arctic and, when appropriate, request the Arctic Council to deliver jointly agreed Arctic messages'; '[The meeting] encourage Arctic states to implement, *as appropriate in their national circumstances*, relevant recommendations for reducing emissions of black carbon (...)' (Arctic Council; 2002a, 2, 2011, 5).

Technical Arctic: '[The Arctic Council] directs SAOs to review the need for an integrated assessment of multiple drivers of Arctic change as a tool for Indigenous Peoples, Arctic residents, governments and industry to prepare for the future, and, based on that review, to make recommendations for consideration.' (Arctic Council, 2011, 6).

Endangered Arctic: 'Noting with concern the findings of the United Nations Intergovernmental Panel on Climate Change that the Arctic will continue to warm at twice the rate of the global average, increasing the likelihood of severe impacts in the Arctic and around the world, and recognizing that ambitious global and national action is needed to reduce climate risks, increase prospects for effective adaptation, and reduce the costs and challenges of mitigation in the long term.' (Arctic Council, 2015, 4).

We composed these worldviews based on our analytical framework, that is by starting from the 'translation outcome' and working back until we reached the content to be translated. Therefore, the worldviews found within the ministerial declarations represent a blueprint for the analysis of SAOs and working groups' translated worldviews, whereas absence from the ministerial declarations would indicate that the worldviews in question failed to cross the obligatory passage points. During the translation process, the different working groups, SAOs, ministers, and secretariats collectively undertake drafting and reviewing efforts until a consensual text is arrived at. Throughout this analytical exercise, we observed how the scientific idiom gives place to a political one. Moreover, this analysis enabled us to observe variations in the gradation of worldviews between ministers, SAOs, and working groups, and between the different working groups themselves. In the latter, some worldviews are more present and more often expressed than others, configuring a sense of group identity within the different actors of the Arctic Council.

Although Machen (2018) points to the importance of power dynamics in translation, there are multiple variables that could explain why one worldview prevails over another: from the individual characteristics of the translator (gender, age, nationality, personality of the 'translator', historical moment, and so on) to intersubjective factors, such as the coherence or otherwise of a specific worldview for a different group of actors, and even the compatibility or otherwise of different worldviews. Identifying which worldview prevails within a group, and how this same worldview is transformed or erased by others, allows insights into the overall science–politics dynamics of the Arctic Council. In the next section, we detail the concept analysis for SAO reports and for each working group's progress reports. We highlight how the worldviews vary according to the group that utters them. We observe variations not only in the degree to which worldviews are present, but also in how they are presented.

5.4 Crafting a Report: Translation and the Lexicographic Role of a Boundary Object

Within the Arctic Council, the translation from a scientific to a political idiom takes place at the obligatory passage points through boundary objects. As boundary objects, reports connect these different realms together. The two points of juxtaposition between science and politics are represented by SAOs' summary reports to ministers and the progress reports from working groups to SAOs. If translation covers meaning, form, function, intention and outcome (Machen, 2018), Arctic Council authors have specific intentions when drafting their reports, trying to preserve a meaning while catering for a specific audience. For instance, when working group assessment reports read as highly technical, they are transformed into a more general and adjectivised style.[8] During the meetings and throughout the exchanges of drafts, different interpretations are articulated and tensions between scientific and political claims are negotiated, and taken care of, among the different authors. Therefore, based on the worldviews identified in ministerial declarations, we move to the analysis of the SAO reports to the ministers, observing how different worldviews move across passage points, travelling along the science–politics spectrum.

5.4.1 SAO Reports to Ministers

SAO reports to Minister are considered to be the baseline documents on which ministers draft their declarations (Rottem, 2020). They are drafted during SAO executive meetings, which are considered to be very strict gatherings where 'real discussions' take place (Spence, 2017). On the other hand, those reports are compiled from information derived from progress reports provided by the working groups and presented by their respective Chairs at the SAOs meetings twice a year. Once again, as a report travels across increasingly political levels, it becomes more and more necessary for the SAOs to maintain the scientific conclusions originally drafted by the authors. Nevertheless, even though ministerial declarations have been considered an endorsement of SAO reports (SAOs would say they are 'writing for themselves'), there are significant differences between them regarding their length and conceptual choices (see Fig. 5.5).

When we compare the word frequency and conceptual cloud of SAOs reports to that of ministerial declarations, the most striking difference is that concepts such as 'cooperation', 'people' and 'government' are no longer at the centre nor ranking as high, whereas the concepts 'project', 'activity', 'assessment', and 'implementation' become far more prominent. Another meaningful point is that concepts like

[8] By 'adjectivise' we mean the transformation of nouns into words whose function is the provision of 'qualities', 'descriptions' ascribable to nouns. A highly adjectivised text will have more of a descriptive than an 'active' character.

Concepts	Count
Project	2390
Activity	1668
Work	1580
Development	930
Information	805
Plan	664
Implementation	618
Cooperation	609
Assessment	605
Change	577
Recommendation	551
Community	543
Arctic States	517
Climate	516
Action	477

Fig. 5.5 SAOs report to Ministers concept cloud from 1998 to 2021

'challenge' and 'concern' were only present at the ministerial level, whereas concepts like 'funding', 'outreach', and 'datum' are only present at SAO level. These differences have important implications when we consider the worldviews translated from level to level. While a *cooperative Arctic* worldview might have been preeminent at the most political level, it is gradually being replaced by the *technical Arctic* worldview which shifts to the centre of communication at SAO level. Of course, this shift can be associated with the intention behind SAO reports: to inform ministers about the work of working groups, which is in essence technical. On the other hand, the intention of ministerial declarations is to share a message with a myriad stakeholders, especially their domestic audiences. Likewise, the almost complete absence of the *endangered Arctic* worldview in SAO reports is another indicator that the creation of a sense of urgency is a role performed much more by the most political actors (and much less by the most scientific ones).

In procedural terms, SAOs want their reports to ministers to contain a minimum level of detail from working group reports. Structurally, they should contain a summary of each working group's deliverables, followed by SAOs' recommendations to ministers. These recommendations will form the basis for the preambles to and operative paragraphs of ministerial declarations. As an illustration of the simplification process that information undergoes, we took the SAO report to ministers in April 2009, which informs the latter about the achievements of the International Polar Year that took place during the Norwegian Chairmanship. Under the main chapter heading '3. Overview of Activities during the Norwegian Chairmanship', several subheadings are listed such as 'integrated resource management', 'climate change', and 'thematic activities' (Arctic Council 2009). In this chapter, the 'International Polar Year' is connected with sixteen specific recommendations by SAOs. In the subsequent ministerial declaration, nine of these recommendations remain in place. Whereas some recommendations are literally 'copied and pasted' such as calls for national funding consultations, or for support for continued international coordination, other specific recommendations such as that the Association of Polar Early Career Scientists (APECS) be recognised have simply disappeared.

To understand these differences between the ministerial and the SAO level in terms of reporting, we also need to consider the information on which SAOs reports are based: the working group progress reports.

5.4.2 *Working Group Reports in Senior Arctic Officials' Executive Meetings*

As we focus our analysis on working groups, we are confronted by an abundance of reports defining the contribution of the Arctic Council's work throughout the years. Although on first sight we might assume that working groups operate on similar terms, a closer look reveals subtle differences between them, a reflection of their different origins and mandates. Under the Arctic Environmental Protection Strategy (AEPS) umbrella, AMAP, CAFF, EPPR and PAME were the first working groups created to address urgent environmental issues in the Arctic that required cooperation. Initially responsible for monitoring pollution, AMAP has been the group with the strongest scientific and technical foundation since its inception. Throughout the years, monitoring and assessment projects shifted more towards climate change issues, corroborated by reports such as the Arctic Climate Impact Assessment (ACIA). Thus, for example, addressing the conservation of flora and fauna, CAFF moved from being a more policy-oriented to a more technical working group, monitoring and assessing the 'health of Arctic ecosystems', especially after the attempt to create a circumpolar protected areas network came to a halt in 2004 (Rottem, 2020, 296). This shift is evidenced by the Arctic Biodiversity Assessment (ABA) report from CAFF, published in 2013.

By contrast, the PAME and EPPR working groups kept their policy-oriented profile, focusing on identifying the main challenges to Arctic shipping and recommending best practice for maritime safety in the region. Although the Arctic Council does not produce legally binding instruments, discussions taking place within working groups informed agreements outside the Arctic Council. For instance, the Arctic Marine Shipping Assessment (AMSA) produced by PAME in 2009 informed the International Maritime Organisation and the negotiations that led to the framing of the Polar Code (Barry et al., 2020, 6). Likewise, EPPR also had an important impact on Arctic policy through the introduction of two legally binding instruments: the Agreement on Cooperation on Aeronautical and Maritime Search and Rescue in the Arctic (SAR Agreement) in 2011, and the Agreement on Cooperation on Marine Oil Pollution Preparedness and Response in the Arctic (MOSPA Agreement) in 2013. Nevertheless, some critics note that these agreements did not create new obligations, having been built from pre-existent commitments (Barry et al., 2020, 4).

Therefore, the real policymaking capacity of PAME (and of the Arctic Council in general) needs to be qualified in this respect.

These differences between the working groups are matched by similar variations in the construction of their reports. For instance, the assessment work undertaken by AMAP, CAFF and PAME (to some extent) leads to a far greater use of technical language in their progress reports, with the *technical Arctic* worldview prevailing. The prevalence of the technical in their work combined with the longevity of these groups (in operation since 1991) is quite well illustrated by the amount of detail they share about their reporting procedures, both in terms of drafting and revision. Within AMAP, different scientific standpoints are synthetised into a single type of document called a 'scientific assessment report'. The secretariat initiates the process by circulating a draft version to national contact points while an independent peer review takes place. Once a definitive version is agreed upon, an editor prepares an outline for the summary for policymakers, based on consultations with the leading authors. The main goal is to 'distil' the key scientific messages that will constitute the summary. From this moment on, an iterative process takes place between the editor and leading authors, in which the latter comment on the editor's proposed draft. Once a draft is ready, this document is opened for comments by permanent participants and heads of delegation, until tacit approval is reached (or the absence of further comments on the draft version).

By repeating the same analysis of AMAP's progress reports from 1998 to 2021, we ended up with a very different concept cloud compared to those for the SAOs' reports and ministerial declarations (see Fig. 5.6). There are far more references to the work that AMAP has undertaken such as 'assessment', 'monitoring', 'work', 'expert', while their object of assessment is also clearly represented by terms such as 'climate', 'POP(s)' (persistent organic pollutants), and 'mercury', so that we can

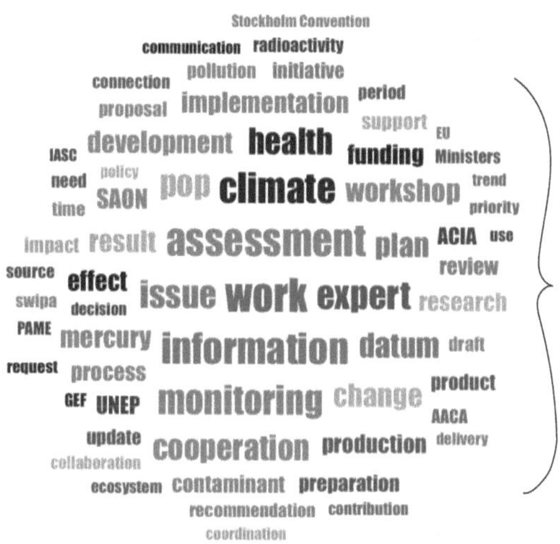

Concepts	Count
Work	921
Assessment	850
Information	349
Climate	333
Monitoring	325
Expert	298
Issue	263
POP (Persistent organic pollutant)	257
Health	252
Cooperation	246
Datum	244
Plan	241
Change	227
Workshop	205
Result	188

Fig. 5.6 Concept cloud for AMAP progress reports 1998–2021

conclude quite straightforwardly that the *technical Arctic* worldview is prevalent within AMAP. On the other hand, we also identified references to 'cooperation', 'coordination', and 'collaboration', which reflect a *cooperative Arctic* worldview. Corroborating this working group's more collaborative nature, the concept cloud also shows references to other bodies and instruments, such as the 'Stockholm Convention', the 'United Nations Environmental Program (UNEP)', the 'Sustaining Arctic Observing Networks (SAON)', and the 'European Union (EU)'. It is also quite enlightening to note the absence of references to other worldviews such as *national Arctic*, *human Arctic*, and *endangered Arctic*, which are present in the reports of other working groups.

Following similar lines to AMAP, CAFF provides details of its internal organisation and its main reporting practices. CAFF's main structural element is a Management Board, composed of a chair (generally from the host country of CAFF's annual meetings), two vicechairs (the outgoing and upcoming chairs), and representatives of the eight Arctic states. The Management Board is assisted by the Executive-Secretariat. The CAFF International Working Group, however, includes representatives not only of the member states, but also of Indigenous peoples' organisations as well as permanent and *ad hoc* CAFF observers. CAFF progress reports are produced annually for the attention of SAOs and biannually for that of ministers through the SAOs. Internally, lead countries present the progress of their projects and subgroups present their findings and recommendations to the International Working Group. Chairs and the secretariat also report to this group annually, and twice a year to the national representatives on the Management Board (CAFF, 1992).

As far as writing the progress reports is concerned, the secretariat, with the support of rapporteurs, prepares draft reports of CAFF's working group meetings, which are distributed before the end of the sessions. The reports contain information on the discussions around each agenda item and a summary of all agreements reached. It also presents the progress of CAFF's workplan and a consideration of SAOs' and ministers' tasks. In addition, there is a record of discussions on issues not previously perceived as common concerns, the rationale supporting proposals, and recommendations to Ministers and SAOs with regard to CAFF's workplans. The final report needs to be agreed by the eight Arctic states (CAFF, 1992). Again, as in AMAP's concept cloud, CAFF's main concepts are related to their specific mandate such as 'monitoring', 'biodiversity', 'ecosystem', 'specie(s)', 'seabird', 'population' and 'datum', which again indicates the prevalence of the *technical Arctic* worldview.[9] But in contrast to AMAP, the *human Arctic* worldview is far more present here in concepts such as 'development', 'people', and 'community'. If we compare AMAP and CAFF's main worldviews, although both give a highly

[9]The appearance of 'specie' in the word cloud results from the software's stemming and corpus-processing functions rather than a conceptual categorisation (that is, it does not actually refer to the real meaning of 'specie' as contrasted to 'species' in English), and does not affect the interpretation of the results.

Concepts	Count
Monitoring	800
Ecosystem	467
Biodiversity	458
Specie	389
Change	382
Datum	375
Development	355
Implementation	294
Effort	267
CBMP (Circumpolar Biodiversity Monitoring Programme)	260
Workshop	257
Management	240
Information	230
Population	230
Program	227

Fig. 5.7 Concept cloud for CAFF progress reports 1998–2021

technical representation of the Arctic, the one gravitates more towards institutional cooperation, whereas the other leans towards the Arctic's local populations.

Focused on the protection and sustainable use of the Arctic's marine environment, PAME is far more policy-oriented than the two previous working groups (see Fig. 5.7). Its main outputs are guidelines, best practices, and reports, although assessments are also part of its work. The members gather twice a year to assess the progress of the group's work and to discuss the development of their workplans. Structurally, PAME places itself as a coordinator between SAOs and expert groups (Arctic Shipping, Marine Protected Areas, Ecosystem Approach to Management, Resource Exploration and Development, and Marine Litter in the Arctic), presenting itself as a communicator of tasks originated by SAOs to the expert groups whilst compiling the results of their projects as findings to be presented to the SAOs. There is not much available information about how PAME's progress reports are drafted and reviewed, but we can observe similar patterns to AMAP and CAFF's concept clouds.

PAME's progress reports are marked by a substantial presence of references to the group's work designated by terms such as 'information', 'progress', 'effort', 'recommendation' and 'expert' (see Fig. 5.8). Likewise, there is also a significant occurrence of terms specific to its mandate such as 'oil', 'shipping', 'development', 'gas', 'ecosystem', and 'regional action plan' (RPA). But like AMAP (and unlike CAFF), there are references of 'collaboration' and 'cooperation', but not to 'people' or 'population'. While we might initially conclude that PAME is embedded in a *technical Arctic* worldview, the *cooperative Arctic* seems to be prevalent as well. Considering that we previously defined references to 'development' as related to a *human Arctic* worldview, because ministerial declarations link economic development with benefits to Arctic peoples 'harnessing the economic potential of the region for the benefit of Arctic inhabitants' (Arctic Council, 2019, 16), in PAME's case, this association is not that straightforward. Analysis of PAMEs' concept cloud shows that 'development' is directly related to resource exploration, rather than to the concepts 'peoples', 'population', and 'Indigenous', as they do not score highly.

Concepts	Count
Work	440
Development	271
Information	271
Draft	214
Plan	203
Shipping	202
Implementation	201
Ecosystem	197
Expert	190
Workshop	173
RPA (Regional Programme of Action)	172
Progress	165
Management	148
Recommendation	143
EA (Ecosystem Approach)	140

Fig. 5.8 Concept cloud for PAME progress reports 1998–2021

As our goal is to observe how worldviews have been transmitted from one social world to the other, we might conclude that 'development' gets connected to the idea of 'Arctic populations' in more political realms. Whereas PAME's representation of 'development' is far more connected to a 'resource-rich Arctic', a worldview which is not present either at the SAO or at ministerial level.

With a strong focus on collaboration for the improvement of the capacities and capabilities to respond to human and environmental emergencies in the Arctic, EPPR is essentially a policy-oriented working group. Its work is based on expanding the understanding of Arctic emergencies while developing guidance and risk assessment tools and coordinating Arctic response exercises. It also sets itself up as the maintainer of MOSPA and SAR agreements, a quite different role from the other working groups. EPPR also meets twice a year to share information and discuss the progress of the projects listed in the workplan and managed by its expert groups (Search and Rescue, Marine Environmental Response, and Radiation). Like PAME, EPPR does not detail how its reports are drafted and reviewed by the group, but we can still identify its main worldviews by analysing the progress reports. Although one might question the relevance of detailing the procedures of some working groups' drafting and not others', it is quite enlightening to consider that the most scientific and experienced groups are precisely those that make such procedures available. The systematisation of report drafting reinforces the concern of these groups for the clarity and objectivity of the information that they convey.

EPPR's concept cloud (Fig. 5.9) shows how progress reports are almost an exact reflection of its mandate: 'response', 'oil', 'information', risk', 'spill', 'emergency', 'prevention', 'exercise', and 'preparedness' are the main significant concepts. These are far more related to the *endangered Arctic* worldview, which seems to be prevalent within this working group. The terms 'disaster' and 'accident' are referred to directly, whereas concepts like 'assessment', 'analysis' and 'mapping' (central to the science-oriented working groups) are less prominent within EPPR. But we can

Concepts	Count
Response	381
Information	305
Oil	305
Emergency	293
Spill	239
Work	239
Exercise	234
Development	195
Risk	187
System	159
Preparedness	130
Prevention	130
Recommendation	129
Plan	121
Proposal	118

Fig. 5.9 Concept cloud for EPPR progress reports 1998–2021

identify other worldviews: there are references to 'community' and to 'cooperation' in EPPR's concept cloud. The presence of the *human Arctic* and *cooperative Arctic* worldviews is coherent with EPPR's mandate, as dealing with human and environmental emergencies relies heavily on cooperation between the different Arctic stakeholders, and particularly with Arctic communities.

Whereas AMAP, CAFF, PAME and EPPR were the first working groups that were originally part of AEPS before their incorporation to the Arctic Council, ACAP and the SDWG were created after the Arctic Council was operative. SDWG was created in 1998 with a focus on protecting and enhancing the environment and the economies, culture and health of Indigenous peoples and Arctic communities. Its mandate is rooted in providing practical knowledge and capacity building to these groups. Arctic states and Indigenous peoples participate directly together with accredited observers. It has two subsidiary expert groups (on Arctic Human Health and on Social, Economic and Cultural aspects), but there is no further information on its internal organisation, or on its standard procedures for drafting and reviewing progress reports. Nonetheless, the analysis of SDWG progress reports to SAOs provides significant insights into the group's worldviews and how different it is from the other working groups.

With its focus on the human dimension in the Arctic, the SDWG concept cloud (Fig. 5.10) has 'development' at its centre. Other terms associated with the *human Arctic* worldview such as 'community', 'participant', 'people' and 'youth' are equally present. In no other forum within the Arctic Council, is the *human Arctic* worldview so strongly represented. As with any other working group, we also identified concepts connected to their practical work, such as 'information', 'expert, 'discussion', 'proposal' and 'recommendation', which shows that the *technical*

Concepts	**Count**
Work	419
Development	282
Health	227
Issue	206
Information	175
Community	172
Plan	169
Proposal	167
Datum	163
Workshop	157
Discussion	141
Participant	136
Recommendation	130
Energy	124
People	113

Fig. 5.10 Concept cloud for SDWG progress reports 2000–2021

Arctic worldview is also present. There is a reference to 'cooperation', but the *cooperative Arctic* does not seem to be as prevalent as in other working groups.

The last working group to be created, ACAP is focused on the reduction of environmental risks and on the prevention of pollution but through a capacity-building approach. Its work is strongly policy-oriented, as it endeavours to provide solutions to identified problems. Projects within ACAP's expert groups (POPs and Mercury, Waste, Indigenous Peoples' Contaminant Action Program, and Short-lived Climate Pollutants) also highlight the needs of Indigenous peoples and the incorporation of traditional and local knowledge into their work. Although there is not much information available on working group functioning, or on how progress reports are drafted and reviewed, we still can deduce the group's worldview from their progress reports from 2006 to 2021.

ACAP's concept cloud (Fig. 5.11) has similar features to those we identified in EPPR. The main concepts are related to ACAP's mandate: they are technical to some extent, but with undertones that make the *endangered Arctic* worldview dominant. The concepts 'pesticide', 'emission', 'mercury', 'waste', 'source', 'destruction', and 'reduction' occupy the centre. In other words, ACAP's reports are focused on problems the groups want to solve, more than on technical work emptied of normative connotations. For instance, 'pure' technical concepts such as 'implementation', 'demonstration', 'assessment', and even 'analysis', are present, but in secondary positions. As consideration of Indigenous peoples' needs and knowledge is foundational to ACAP, the *human Arctic* worldview is also present through terms such as 'community' and 'participant'. There is also a reference to 'cooperation', but it is considerably weaker than in other working groups.

Concepts	Count
Pesticide	304
Emission	291
Phase	280
Mercury	266
Waste	258
Region	245
Source	240
Destruction	226
Facility	216
Work	214
Implementation	202
Reduction	201
Management	192
PCB (Polychlorinated Biphenyl)	183
NEFCO (Nordic Environment Finance Corporation)	179

Fig. 5.11 Concept cloud for ACAP progress reports 2006–2021

Considering the volume of reports and assessments these different working groups produce, this conceptual analysis would certainly look different if we had included other documents. Nevertheless, as our focus is the interface between science and politics, the analysis of progress reports, SAOs' reports and ministerial declarations gives us a continuum initiated within working groups and finalised at ministerial meetings. Throughout this continuum, worldviews were translated, being shaped by the forums in which they were presented, whilst retaining elements recognisable to their original authors (see the summary Table 5.2). In the next section, we draw conclusions about how these worldviews changed as they travelled through the science–politics spectrum, at the same time as they provided a sense of identity to the different groups that enacted them.

5.5 Conclusion

Taking the ministerial declarations as our blueprint, we have been able to observe retrospectively how different worldviews were translated from levels dominated by science to a level dominated by politics. Within the ministerial level, the worldview *endangered Arctic* is strongly connected to climate change and its impact on livelihoods. This worldview is going to be very significant to the EPPR, but only with regard to the threat presented by human and environmental emergencies. For ACAP, the *endangered Arctic* is also crucial, but its perspective is more focused on the dangers of pollution to the environment and people's health. Curiously, we could not find concepts associated with the *endangered Arctic* at the SAO level. The *cooperative Arctic* worldview is broadly present within the science/politics interface, but to different degrees. Whereas cooperation is pivotal for ministers and quite relevant for SAOs, at the working group level only AMAP would give cooperation such a high degree of prominence. In PAME, this worldview is represented more by the

Table 5.2 Arctic worldviews in Arctic Council reporting

Authors	Main Arctic worldviews	Exemplary quotations
SAOs	Technical Arctic	'Developments outside the Arctic have a strong influence on the preconditions for sustainable development and environmental protection in the Arctic. The AC has an important role in *informing international bodies* about the *results of its work* and with consensus among the Arctic states make [*sic*] efforts to increase *understanding* of Arctic concerns in a global context' (Arctic Council, 2002b, 6)
AMAP	Technical Arctic (with an undertone of Cooperative Arctic)	'AMAP's mission is to monitor and assess the status of the Arctic region with respect to pollution and climate change issues by (i) facilitating and advancing the *coordinated* implementation of relevant circumpolar *monitoring and research*, (ii) documenting levels and trends, pathways and processes, and effects on ecosystems and humans, (iii) distinguishing human-induced changes from changes caused by natural phenomena, and (iv) proposing actions to reduce associated threats *for consideration by governments and relevant organisations'* (AMAP, 2019, 3)
CAFF	Technical Arctic (with an undertone of Human Arctic)	'In order to successfully conserve the natural environment and allow for *economic development baseline data* and status and trend analyses of Arctic biodiversity, habitats and ecosystem health are required. CAFF's projects *provide data for informed decision-making* in resolving conflicts which are now arising in trying to both conserve the natural environment and permit regional growth' (CAFF, 2010, 1)
PAME	Human Arctic (with an undertone of resource-rich Arctic)	'*Abundant natural resources*, increasing transportation and *economic activity*, and significant changes due to climatic processes, are resulting in increased use and threats to the Arctic marine environment. Measures to control and reduce these pressures and impacts exist but they have largely been reactive and developed on a sector-by-sector basis resulting in a patchwork of policies, legislation and programmes. [PAME was tasked to] [d]evelop an Arctic Council strategic plan for the protection of the Arctic marine environment based on comprehensive, collaborative and integrated *ecosystem management approaches'* (PAME, 2003, 1)
EPPR	Endangered Arctic (with an undertone of Technical Arctic)	'The *harsh conditions* and the lack of existing infrastructure in much of the Arctic make *emergencies more problematic* than in more temperate climates. Actions for prevention and preparedness must be adapted to the harsh conditions and *remoteness* of the Arctic. *Research* is being done at universities and at industrial and governmental research centres to *further develop this knowledge*. International cooperation in this area is of major importance to accelerate programmes; maximise resources, and find solutions to *common problems'* (EPPR, 2010, 27)

(continued)

Table 5.2 (continued)

Authors	Main Arctic worldviews	Exemplary quotations
SDWG	Human Arctic	'The Arctic region is the *home* of approximately four million *people*. A sustainable economic growth is important both for the environment and livelihood of *Arctic residents*. The Arctic region is unique with its extreme climate and sensitive nature. At the same time, it is *rich in natural resources* such as flora, fauna, energy and mineral deposits. The melting ice is also providing *new opportunities for transport and logistics*. If the Arctic is to be developed in a sustainable manner, it must attract responsible companies that will play a key role in the *economic development* of the region and beyond' (SDWG, 2014, 2)
ACAP	Endangered Arctic (with an undertone of Human Arctic)	'ACAP addresses *Arctic pollution sources* and acts as a strengthening and supporting mechanism to encourage national actions to *reduce emissions* and other *releases of pollutants* that are relevant in the Arctic. Cooperative actions make an important and significant contribution to the overall international effort to *reduce environmental damage* on a global level. ACAP will develop concrete project proposals within this mandate for approval as Arctic Council projects, taking into account the needs of *Indigenous populations* in the Arctic' (ACAP, 2015, 5)

concept of 'collaboration'. The *human Arctic* worldview offers another angle from which to observe the different ways that translation can take place at the Council. Strongly present at the ministerial level, and (to a lesser extent) among SAOs, this worldview is not present in AMAP's concept cloud and only slightly in ACAP, EPPR, and CAFF's. Conversely, the *human Arctic* worldview defines PAME and SDWG group identity, but in different ways. Both have 'development' as their core concept, but PAME approaches it through economic development, whereas SDWG is centred on protecting the livelihoods of Arctic communities.

More striking than these worldviews are the results for *national Arctic* and *technical Arctic*. The *national Arctic* worldview is strongly represented in the concept clouds at the ministerial and SAO levels but not in the working groups, whereas the *technical Arctic* is present in all working groups' concept clouds. However, as we move up the science–politics continuum, the technical terminology fades away. In the table below, we juxtapose these different worldviews. As we have just described, some worldviews get lost whilst others become prevalent through the different translations in the passage points. Reports, as boundary objects, uphold the identities of the different groups, keeping science and politics as distinct realms. Notwithstanding the maintenance of these boundaries, however, reports also intertwine science and politics in transmitting worldviews. For instance, the *endangered Arctic* which is a defining worldview in EPPR and ACAP through the concepts of 'disaster' and 'destruction' respectively, is translated into and represented as a 'challenge' by ministers.

To understand how the boundaries of the science/politics interface operate within the Arctic Council, we focused on observing how different Arctic worldviews travel

Table 5.3 Worldviews in Chap. 5

Worldviews of Ice	Case Study 3: 'Bringing Science to Politics'
Endangered Arctic	
Diagnosis	Contamination, accidents, biodiversity loss and climate change
Solution	Monitoring, assessment and coordination
Epistemological foundation	Evidence-based policy
Cooperative Arctic	
Diagnosis	Latent political rifts and extreme environment
Solution	Fostering relationships and information sharing
Epistemological foundation	Collective action
Human Arctic	
Diagnosis	Economic development and culture protection
Solution	Inclusion of economic actors and local communities
Epistemological foundation	Democracy (participation, representation and deliberation)
National Arctic	
Diagnosis	Challenge to Arctic state's sovereignty
Solution	Reification of national authority
Epistemological foundation	Westphalia system
Technical Arctic	
Diagnosis	Necessity of better understanding Arctic systems
Solution	Funded research
Epistemological foundation	Scientism

through working groups, SAOs and Ministerial Meetings, enduring, disappearing or being transformed as they are translated from one level to another. For this, our analysis took a step back and looked to the reports from 1998 to 2021 as a whole—revealing a structuring group identity that permeates the years of existence and operation of the Arctic Council. But one could question: do Arctic worldviews change over time? They certainly do. The 'technicalisation' of CAFF is one example of how groups can shift through time and, consequently, their worldviews. However, to pinpoint every single fine variation or even every potential trigger of such fluctuations, it would unavoidably lead to a trade-off: as we gain nuance, we lose perspective. Worldviews resonate with their institutional context—therefore, as long as working groups, SAOs and ministers' mandates did not suffer substantive institutional modifications, we were able to distil their worldviews from their uninterrupted reporting periods as a whole. And worldviews have consequences: they are used by actors according to their particular purposes, and these purposes within the Arctic Council have been keeping science and politics boundaries intact (see summary Table 5.3).

Chapter 6
Conclusion

Abstract The concluding chapter summarizes the preceding chapters and shows that and how worldviews matter in the various phases of the production and transmission of Arctic knowledge.

Keywords Uncertainty · Politics of Worldviews

On August 18, 2019, researchers from Rice University in Houston joined members of the Icelandic Hiking Society and the general public in commemorating the former Okjökull glacier in Borgarfjörður, west-central Iceland. The once dome-shaped glacier called 'Ok' had melted away through the course of the twentieth century and been declared dead by glaciologists in 2014. The group gathering atop the Ok volcano installed a memorial plaque with a 'Letter to the future' in Icelandic and English, reading as follows: 'Ok is the first Icelandic glacier to lose its status as a glacier. In the next 200 years all our glaciers are expected to follow the same path. This monument is to acknowledge that we know what is happening and what needs to be done. Only you know if we did it.'[1] The story of the melted glacier was also prominently told by the former mayor of Reykjavik, Jón Gnarr, in the short movie 'Not OK', produced by Rice University anthropologists Dominic Boyer and Cymene Howe in 2018.[2]

In this book, our aim has been to trace the manifold narratives bringing science and politics together in their attempts to grasp the Arctic. Our basic assumption was that these worldviews are far from being strategic devices used by geopolitical interests, or scientific stories intended to speak 'truth to power', or 'mise-en-scènes' hiding the power plays of elites. Rather, we conceptualised worldviews as constitutive for communication and coordination in both Arctic research and geopolitics. In making present a future that is fundamentally uncertain, they manifest themselves in expectations, beliefs, values and narratives about what is relevant for present and future actions, what the issues are in knowledge- and decision-making, and how we

[1] https://news2.rice.edu/2019/07/18/lost-glacier-to-be-honored-with-memorial-monument/ [accessed 02 Nov 2025].

[2] https://www.notokmovie.com [accessed 02 Nov 2025].

M. Albert et al., *The Worldviews of Ice*, Springer Polar Sciences,
https://doi.org/10.1007/978-3-032-25945-5_6

could carry on. In projecting current expectations onto a distant future, they form plausible fictions that delineate the horizons of possibilities in the present.

The story of 'Ok' is closely associated with the worldview of the *endangered Arctic* that runs across our three country cases, Canada, Norway and Germany, as well as throughout the 'full circle' interlinking science and politics: first, in relation to the allocation of research funding; second, in relation to the practices of scientists in permafrost research and glaciology; and third in the translations and passage points characterising science advice at the Arctic Council. By this point, it has probably become clear that the full circle itself is an idealisation. Instead, we have encountered contradictory constellations of institutions and funding goals, ambiguities and ambivalences blurring the seemingly crystal-clear picture of basic research, and an advisory process during which worldviews are transformed and sometimes get lost.

The literature on worldviews has been far from precise about how to identify them, how to capture their internal coherence and how to separate them from each other. We therefore proposed a threefold heuristic of worldviews to structure the comparisons and case studies in the past three chapters. Drawing on the conceptually matured and empirically rich tradition of research on frames, we proposed that worldviews could be disentangled by exploring the ways in which they (1) diagnose geopolitical problems and their causes, (2) formulate expectations about adequate solutions and their implementation in the future, and (3) make evaluative and epistemological claims about the validity and legitimacy of both expectations and actions. We assumed that the ways in which these three 'tasks' are related to each other within as well as between potentially overlapping (or competing) worldviews both motivate and constrain the interactions between Arctic research and politics, mobilise understandings of how to analyse and order world relations, and ultimately lead to the collective formation of the Arctic as a geopolitical and scientific object.

We quickly found out, however, that some worldviews collide with one another, that they sometimes overlap or compete, that they are far from cleanly delineable and rarely set the tone for decisions and deliberations in an explicit manner. To capture the multifaceted picture emerging from these observations, we based our analysis on a rich ground of empirical data. Expert interviews allowed us to explore the motivations and views of actors across multiple levels; bibliometric calculations made it possible to trace the disciplinary shifts characterising Arctic research in a comparative perspective; qualitative studies of policy briefs, strategy papers, media reports, personal blogs and journal articles allowed us to scrutinise the deep entanglements between knowledge-production and policymaking.

All in all, we think that the results of our endeavours go far beyond the recurring stereotypes of a unitary 'voice of science' and its seemingly clear separation from politics. Science funding, research activities and policy advice are intertwined with multiple but not unlimited worldviews shaping a landscape that is of geopolitical and epistemic relevance at the same time. These worldviews are driven by deep uncertainties, 'unknown unknowns', which reappear even after seemingly having been domesticated by scientific risk calculations and predictions (Katzenstein, 2022).

6.1 Messages to Uncertain Futures

The memorial plaque commemorating the death of 'Ok' as a glacier is not the only attempt to send messages to the future. Another prominent case is that of a bottle discovered in the Canadian Arctic in 2013, buried by glaciologist Paul T. Walker in 1959 under a rock cairn. In his message, Walker simply requested the finder to measure the distance to a nearby ice shelf and to report back to his laboratory in Ohio. Fifty-four years later, researchers from Laval University did as Walker had requested and found that the glacier had shrunk dramatically.[3] A more recent attempt to inform the future has been undertaken by permafrost researchers from the Polish Academy of Sciences Institute of Geophysics in Warsaw in 2017 when they buried a stainless-steel tube preserving a diversity of scientific artefacts near a Polish Polar Station in Hornsund. The capsule is supposed to resurface half a million years from now as consequence of geological uplift (Schiermeier, 2017).

In all these messaging activities, the common factor is the uncertainty of the Arctic's future. We found this to be a deep and constant, sometimes even emotional motive for many researchers. In a world of many uncertainties, the reference to worldviews such as the *endangered Arctic* fulfils an ordering function. As research is structured within this framework of climate change, the need for 'evidence-based policy' and more practical knowledge to reconstruct and mitigate the consequences of both climate change and decreasing biodiversity comes to the fore. Knowing the importance of Arctic regions for the global climate system means that any 'unknown unknowns' need to be brought to light and understood all the more urgently. Authorised by this sense of urgency, prominent research gaps are formulated in national and international research programmes or discussed in scientific publications.

Still, there seems to be a tension between the worldviews of the *endangered Arctic* and the *unknown Arctic.* As both Chaps. 3 and 4 show, the *unknown Arctic* normally speaks to and supports the *endangered* worldview of the Arctic. Sometimes, however, the unknown stands in conflict with the *endangered* view, because scientific research is limited to working on the most pressing problems in terms of its allocation of resources, staff and funding. This is apparent in national research priorities, too, where the framing of 'unknown unknowns' is closely interlinked with climate change and the resulting applied-research-related frameworks. Ultimately, both worldviews appeal to similar but slightly different futures. While the worldview of the *endangered Arctic* is strongly bound up with the overarching goal of mitigating the consequences of worldwide climatic shifts, the worldview of the Arctic as unknown highlights the self-sufficiency of science and the need for basic research even where it is not directly related to the climate catastrophe. Not only do we find a plurality of worldviews motivating the politico-epistemic quest for knowledge in various ways, more importantly, letting these inherently colliding worldviews shape what seems to be collectively relevant and unquestionable may result in divergent action orientations appealing to different futures.

[3] https://www.oxy.edu/magazine/issues/spring-2014/frozen-time [accessed 02 Nov 2025].

It seems that both politics and science are potentially torn between multiple futures and horizons for further action. This is especially prevalent under conditions of 'deep time uncertainties' (Straßheim, 2021). Deep time refers to the observations of human impacts on planetary processes and the shift from the Holocene to the recently proposed geological epoch of the Anthropocene, in which the states of Earth systems are profoundly altered by human actions and, in turn, impact the living conditions of societies. From this perspective, the problem of deep time uncertainties is not so much the correct calculation of geological timescales. Perhaps more important is the interplay between 'deep time' and 'political time', between the trajectories of Arctic geo- and ecosystems and the timescales of Arctic policymaking. Against this background and given that preparations for the Fifth International Polar Year (IPY-5) in 2032–33 have already begun, it could be helpful to compare politico-epistemic worldviews evoking different futures, to distinguish more explicitly between probable, wanted and unwanted futures for the Arctic at the interfaces of science and policy, and to develop ways of communicating these collaborative efforts within Arctic communities and beyond.

Probably the most sobering result of our research is related to 'indigenous worldviews' (Hart, 2010). Traditional ecological knowledge (TEK) certainly has become a regular point of reference, for example within the worldview of the endangered Arctic. Many TEK holders have experiences working in close cooperation with Western scientists. There is also a growing recognition that Western academics need to go beyond notions of simply 'integrating' TEK into Western science. This would require Western academics and TEK holders to define funding criteria in a joint endeavour, to collectively develop research questions and methodologies, and to co-produce advisory knowledge at the science-politics interface (Rosales & Chapman, 2015). However, we did not come across any explicit reference to broader, more encompassing indigenous worldviews, their epistemic principles and underlying values. Of course, we cannot rule out that such an influence exists. But getting a broader picture of Indigenous peoples' worldviews and their politico-epistemic role would exceed the limits of our study.

6.2 Worldviews Matter

Probably one of the most significant results of our study is that worldviews matter. This is apparent across all levels: For the central Arctic science funding systems in our case-study country, worldviews play an important role in how institutional configurations are shaped and funding organisations prioritise different funding practices. Preparing for Arctic science through funding is not simply a technical procedure, it needs to be understood as a dynamic social process within varying institutional landscapes and established research funding systems. Worldviews can provide influential ideal frameworks in these processes to bridge different perceptions in science and policy and across countries. However, these shared worldviews are also reinterpreted and specified in national or global contexts through various intervening ideas and beliefs leading to worldviews that we summarised as more *national Arctic*- or, by contrast, more *global Arctic*-oriented. The main point was to ascertain not that these differences exist, but how they are reflected in funding practices and in the ways Arctic science is embedded in public research funding systems. All in all, these worldviews are not mere projections of the geopolitical positions and strategic interests held by influential actors. Instead, they have emerged historically, have been embedded in institutional funding constellations and, over time, have selectively reinforced certain funding practices—thus shaping the very foundations of what Arctic science actually is and why it still matters for societies today.

Opening up the black box of Arctic science, we find that worldviews also affect individual researchers' practices when doing permafrost research and glaciology. Four worldviews of ice could be traced: the *endangered Arctic*, *unknown Arctic*, *exceptional Arctic* and *Arctic adventure*. These worldviews serve to underpin the role-perceptions of scientists, direct their attention to epistemic problems and solutions and are meaningful in terms of how research on ice is perceived in the Arctic. Arguably, the *endangered Arctic* is the most prominent idea, influencing other current conceptions. Climate change is perceived as one of the most pressing geopolitical issues, and the solution that emerges from this diagnosis consists in generating more knowledge. Here, the need for evidence-based policy serves as the epistemological foundation and legitimation for policy advice.

Interestingly—and as already shown above—the *endangered Arctic* is understood much more narrowly with reference to climate change. By contrast, looking at these regions as the *unknown Arctic* leaves room for considerations regarding basic uncertainties in research and the exploration of solutions in basic as opposed to applied research. Here, the basic foundations of science—science as curiosity-driven, science for the sake of science—serve as the epistemological underlay. *Arctic adventure* speaks to the expectations and motivations of researchers when examining the reasons behind their research from an individual point of view. Field work experience, 'being there', provides scientists with expertise that distinguishes them from others who do not go north. Storytelling is used as successful tool to promote and communicate interests—whether scientific, political, societal or personal. Lastly, the *exceptional Arctic* attests a geopolitical crisis in which science is perceived as a politically neutral endeavour, promoting international scientific cooperation and science diplomacy. By distinguishing these worldviews permeating Arctic research, it also becomes clear how they shed light on the science/politics interface and influence the role perceptions of researchers. This is clearly also true for the worldviews interpenetrating the communication processes at the Arctic Council.

6.3 The Politics of Worldviews

To be sure, worldviews are a strategic resource of geopolitical interests. They are taken up and disseminated by science-policy entrepreneurs who use them to defend their strategic positions. In this sense, worldviews are 'weapons' in political struggles. They come into play when research funding is funnelled, when funding decisions are justified across multiple institutions and when philanthropists intervene. Arctic researchers draw on worldviews when they formulate research designs and deal with the radical uncertainties they are confronted with. And in the Arctic Council they are actively transformed by Senior Arctic Officials. SAOs influence the diffusion and translation of worldviews.

In research funding systems and in Arctic research into glaciology or the permafrost, philanthropic organisations are playing an increasingly strong role by influencing which research directions are seen as desirable and worth continuing with. At the level of the Arctic council, SAOs are boundary actors operating as intermediaries between levels, reporting to ministers on the progress of working groups while also guiding the latter's work according to the priorities of ministerial declarations. Interestingly, SAOs act as 'passage points' on a mesoscale: they play a pivotal role in demarcating the boundaries between Arctic science and politics. Their reporting practices at the passage between science and politics significantly influence both the upholding of the boundaries between science and politics and the passage of certain worldviews across them.

For instance, when drafting reports, scientists need to negotiate and select what they believe is the most important information for policy-makers' consideration. Therefore, politics already permeates this scientific realm, as the inclusion/exclusion of information is an interpretative process embedded in the different worldviews of leading authors. By synthesising the information to be communicated, the reports drafted by working groups and SAOs converge to a specific worldview not only of the Arctic, but also of the science–politics dynamics, forging a specific identity for each group.

At the same time, SAOs need to carefully orchestrate their reports when crossing the boundary and moving from the more scientific to the more political level. Within the Arctic Council, several negotiating stances are adopted within working groups, among SAOs and among Ministers. Working groups negotiate the final versions of their reports and their summaries for policy-makers; SAOs and permanent participants negotiate the summary reports from their annual meetings and their reports to ministers; and ministers negotiate their declarations in their biennial meetings. Some worldviews get lost whilst others become prevalent through different translations in the passage points. Reports, as boundary objects, uphold the identities of the different groups, keeping science and politics as distinct realms.

Notwithstanding the maintenance of these boundaries, reports also intertwine science and politics by transmitting worldviews. For instance, the *endangered Arctic*, which is a defining worldview in the expert group on Emergency Prevention, Preparedness and Response (EPPR) and the Arctic Contaminants Action Program (ACAP) on the basis of the concepts of 'disaster' and 'destruction' respectively, is translated and represented as a 'challenge' by ministers. Even more striking are the transitions affecting the *national Arctic* and *technical Arctic* worldviews. The *national Arctic* worldview is strongly represented in reports at the ministerial and SAO levels but not in the working groups, whereas the *technical Arctic* worldview is present in all working groups. However, as we move up in the science–politics continuum, this technical terminology fades away.

At this point, the full circle closes. We have been able to retrospectively uncover the translation of worldviews from levels dominated by scientists to levels dominated by policy-makers. While we have reconstructed both the institutional sedimentation and activation of worldviews in different funding systems and shed light on the ways worldviews influence role perceptions and epistemic priorities within Arctic research, the Arctic Council allowed us to trace the transition journey of worldviews where science and policy meet.

In times in which the brute pursuit of interests and power seems to dominate Arctic geopolitics, it might be helpful to bear in mind how deeply worldviews are ingrained in the collective memory and the institutional configurations of science-politics interactions. Without any doubt, political dynamics influence how worldviews are communicated. We have, however, also shown that worldviews significantly shape the stability and permeability of the boundaries between science, politics and society. The public responses to the attempts of the Trump administration to annex Greenland, as well as the solidarity protests across Greenland, Denmark and multiple other countries confirm the insight that worldviews have the power to mobilise protests and the articulation of collectively shared values.

The Arctic takes shape in a multitude of worldviews of ice. In the present book, we have traced how these worldviews influence, and are constructed in, various parts of what we have conceived to be a circle between science funding, 'doing' science, and science-policy interaction. These worldviews make for the Arctic as a complex social arrangement. They will change, both in themselves, as well as in their relation to each other. However, they will not go away, even when temporarily displaced by geopolitical tensions.

References

Aagaard, K., Mongeon, P., Ramos-Vielba, I., & Thomas, D. A. (2021). Getting to the bottom of research funding: Acknowledging the complexity of funding dynamics. *PLoS One, 16*(5), e0251488. https://doi.org/10.1371/journal.pone.0251488

ACAP (Arctic Contaminants Action Program). (2015). *ACAP Draft SAO Report to Ministers including work plan*. Retrieved from http://hdl.handle.net/11374/1437

ACAP (Arctic Contaminants Action Program). (2025). *Arctic Contaminants Action Program*. Retrieved January 29, 2026, from https://arctic-council.org/about/working-groups/acap/home/

ACIA (Arctic Climate Impact Assessment). (2005). *Impacts of a warming arctic: Arctic climate impact assessment*. Cambridge University Press.

Aksnes, D. W., & Rørstad, K. (2019). *Norwegian Polar Research, High North Research and research in Svalbard: A mapping survey 2018*. Nordic Institute for Studies in Innovation, Research and Education. Retrieved from https://www.forskningsradet.no/siteassets/publikasjoner/2019/nifureport2019-24_ferdig.pdf

Aksnes, D., Osipov, I., Moskaleva, O., & Kullerud, L. (2016). *Arctic research publication trends: A pilot study*. UArctic. Retrieved from https://www.uarctic.org/media/1596721/arctic-research-publication-trends-august-2016.pdf

Aksnes, D. W., Blöcker, C., Colliander, C., & Nilsson, L. M. (2023). *Arctic research trends: Bibliometrics 2016–2022*. Arctic Centre, Umeå University. Retrieved from https://zenodo.org/records/7961982

Aksnes, D. W., Danell, R., Kullerud, L., & Nilsson, L. M. (2024). *Arctic research trends: External funding 2016–2022*. Arctic Centre, Umeå University. Retrieved from https://umu.diva-portal.org/smash/get/diva2:1828711/FULLTEXT01.pdf

Albert, M. (2024). Imaginaries of and in Svalbard: The "making" of an archipelago. In M. Albert, D. Brode-Roger, & L. Iversen (Eds.), *Svalbard imaginaries: The making of an Arctic archipelago* (pp. 1–16). Palgrave.

Amadae, S. M. (2015). *Prisoners of reason: Game theory and neoliberal political economy*. Cambridge University Press.

AMAP (Arctic Monitoring and Assessment Programme). (2019). *AMAP Strategic Framework 2019+*. Retrieved from https://www.amap.no/documents/doc/amap-strategic-framework-2019/1802

AMAP (Arctic Monitoring and Assessment Programme). (2024). *About*. Retrieved January 29, 2026, from https://www.amap.no/about/

Andreeva, S., & Hønneland, G. (2023). *Arctic research and its actors: A pre-study mapping of research strategies and priorities*. Fridtjof Nansen Institute. Retrieved from https://www.fni.

M. Albert et al., *The Worldviews of Ice*, Springer Polar Sciences, https://doi.org/10.1007/978-3-032-25945-5

no/publications/arctic-research-and-its-actors-a-pre-study-mapping-of-research-strategies-and-priorities

Arctic Council. (1991). *Arctic environmental protection strategy (Rovaniemi Declaration).* June 1991, Rovaniemi, Finland. Retrieved from https://library.arcticportal.org/1542/1/artic_environment.pdf

Arctic Council. (1996). *Ottawa Declaration. Declaration on the establishment of the Arctic Council.* Ottawa, Canada. Retrieved September 19, 1996, from http://hdl.handle.net/11374/85

Arctic Council. (2002a). *Inari Declaration. On the occasion of the Third Ministerial Meeting of the Arctic Council.* Inari, Finland. Retrieved October 10, 2002, from http://hdl.handle.net/11374/88

Arctic Council. (2002b). *Report of Senior Arctic Officials to Arctic Council ministers.* Inari, Finland. Retrieved October 9–10, 2002, from http://hdl.handle.net/11374/1552

Arctic Council. (2006). *Salekhard Declaration. On the occasion of the Tenth Anniversary of the Arctic Council, the Fifth Ministerial Meeting of the Arctic Council.* Salekhard, Russia. Retrieved October 26, 2006, from http://hdl.handle.net/11374/90

Arctic Council. (2011). *Nuuk Declaration. On the occasion of the Seventh Ministerial Meeting of the Arctic Council.* Nuuk, Greenland. Retrieved May 12, 2011, from http://hdl.handle.net/11374/92

Arctic Council. (2013). *Arctic Council rules of procedure.* Retrieved from http://hdl.handle.net/11374/940

Arctic Council. (2015). *Iqaluit Declaration. On the occasion of the Ninth Ministerial Meeting of the Arctic Council.* Iqaluit, Canada. Retrieved April 24, 2015, from http://hdl.handle.net/11374/662

Arctic Council. (2019). *Rovaniemi ministerial statements. On the occasion of the Eleventh Ministerial Meeting of the Arctic Council.* Rovaniemi, Finland. Retrieved May 7, 2019. Retrieved from http://hdl.handle.net/11374/2418

Armitage, D., Berkes, F., Dale, A., Kocho-Schellenberg, E., & Patton, E. (2011). Co-management and the co-production of knowledge: Learning to adapt in Canada's Arctic. *Global Environmental Change, 21*(3), 995–1004. https://doi.org/10.1016/j.gloenvcha.2011.04.006

Auswärtiges Amt. (2013). *Germany's Arctic policy guidelines. Assume responsibility, seize opportunities.* Retrieved from https://www.arctic-office.de/fileadmin/user_upload/www.arctic-office.de/PDF_uploads/Germanys_Arctic_policy_guidelines.pdf

Auswärtiges Amt. (2019). *Germany's Arctic policy guidelines. Assuming responsibility, creating trust, shaping the future.* Retrieved from https://www.auswaertiges-amt.de/resource/blob/2582626/eb0b681be9415118ca87bc8e215c0cf4/arktis-leitlinien-data.pdf

Auswärtiges Amt. (2024). *Germany's Arctic policy guidelines. Germany and the Arctic in the context of the climate crisis and the Zeitenwende.* Retrieved from https://www.auswaertiges-amt.de/resource/blob/2676060/5496910022404f7cf68049f1b10e4d5a/arktis-leitlinien-data.pdf

AWI (Alfred-Wegener-Institute). (2020, December 31). *Das AWI im Überblick Zahlen, Daten und Fakten.* Retrieved January 29, 2026, from https://www.awi.de/fileadmin/user_upload/AWI/Ueber_uns/Publikation/Dateien_Printprodukte/st03_Gesch%C3%A4ftsbericht_2020_Factsheet_DE.pdf

Bailes, A. J. K., & Heininen, L. (2012). *Strategy papers on the Arctic or High North: A comparative study and analysis.* University of Iceland Press.

Barry, T., Davíðsdóttir, B., Einarsson, N., & Young, O. R. (2020). The Arctic Council: An agent of change? *Global Environmental Change, 63*, 102099. https://doi.org/10.1016/j.gloenvcha.2020.102099

Beck, S., & Mahony, M. (2018). The IPCC and the new map of science and politics. *WIREs Climate Change, 9*(6), e547. https://doi.org/10.1002/wcc.547

Beckert, J. (2013). Imagined futures: Fictional expectations in the economy. *Theory and Society, 42*(3), 219–240.

Beckert, J. (2016). *Imagined futures: Fictional expectations and capitalist dynamics.* Harvard University Press.

Berkes, F., & Armitage, D. (2010). Co-management institutions, knowledge, and learning: Adapting to change in the Arctic. *Études/Inuit/Studies, 34*(1), 109–131.

Berkman, P. A., Kullerud, L., Pope, A., Vylegzhanin, A. N., & Young, O. R. (2017). The Arctic Science Agreement propels science diplomacy. *Science, 358*(6363), 596–598.

Bertelsen, R. G. (2020). Science diplomacy and the Arctic. In G. Hoogensen Gjørv, M. Lanteigne, & H. Sam-Aggrey (Eds.), *Routledge handbook of Arctic security* (pp. 234–245). Routledge.

Bloom, E. T. (1999). Establishment of the Arctic Council. *The American Journal of International Law, 93*(3), 712–722.

Blum, S., & Straßheim, H. (2025). Experts and expertise in public policy. In M. van Gerven, C. Rothmayr Allison, & K. Schubert (Eds.), *Encyclopedia of public policy* (pp. 1–8). Springer.

Blyth, M. (2002). *Great transformations: Economic ideas and institutional change in the twentieth century*. Cambridge University Press.

BMBF (Federal Ministry of Education and Research). (2012). *Rapid climate change in the Arctic: Polar research as a global responsibility*. Retrieved from https://www.fona.de/medien/pdf/Rapid_Climate_Change_in_the_Arctic.pdf

BMBF (Federal Ministry of Education and Research). (2015). *Rapid climate change in the Arctic: Polar research as a global responsibility*. Retrieved from https://www.fona.de/medien/pdf/pdf_8rch1v/Schnelle_Veraenderungen_in_der_Arktis.pdf

BMBF (Federal Ministry of Education and Research). (2020). *Mare:N—Coastal, marine and polar research for sustainability*. German Federal Government Research Program. Retrieved from https://www.foerderinfo.bund.de/SharedDocs/Publikationen/de/bmbf/FS/31356_Mare_N_en.pdf?__blob=publicationFile&v=4

Bollman, M. (2022). Frames, fantasies, and culture: Applying and comparing different methodologies for identifying energy imaginaries in American policy discourse. *Energy Research & Social Science, 84*, 102380. https://doi.org/10.1016/j.erss.2021.102380

Braun, D. (1993). Who governs intermediary agencies? Principal–agent relations in research policy-making. *Journal of Public Policy, 13*(2), 135–162. https://doi.org/10.1017/S0143814X00000994

Braun, V., & Clarke, V. (2006). Using thematic analysis in psychology. *Qualitative Research in Psychology, 3*(2), 77–101. https://doi.org/10.1191/1478088706qp063oa

Braun, D., & Guston, D. H. (2003). Principal–agent theory and research policy: An introduction. *Science and Public Policy, 30*(5), 302–308. https://doi.org/10.3152/147154303781780290

CAFF (Conservation of Arctic Flora and Fauna). (1992). *Program for the Conservation of Arctic Flora and Fauna. Framework document*. Retrieved from http://hdl.handle.net/11374/2869

CAFF (Conservation of Arctic Flora and Fauna). (2010). *CAFF progress report. Senior Arctic Officials Meeting*. October 2010, Tórshavn, The Faroe Islands. Retrieved from http://hdl.handle.net/11374/1038

CAFF (Conservation of Arctic Flora and Fauna). (2025). *About CAFF*. Retrieved January 29, 2026, from https://www.caff.is/about/caff/#secretariat/

CAFF (Conservation of Arctic Flora and Fauna). (2026). *Circumpolar protected areas network (CPAN)*. Retrieved January 29, 2026, from https://oaarchive.arctic-council.org/collections/2d798074-3b94-4f6b-a37b-852718c4d65c

Callon, M., & Law, J. (1982). On interests and their transformation: Enrolment and counter-enrolment. *Social Studies of Science, 12*(4), 615–625.

Campbell, J. L., & Pedersen, O. K. (2008). *Knowledge regimes and comparative political economy*. Copenhagen Business School. Retrieved from https://research-api.cbs.dk/ws/portalfiles/portal/59064785/wp_cbp_2008_48.pdf

Castoriadis, C. (1997). *The imaginary institution of society*. Polity Press.

Chadburn, S. E., Burke, E. J., Cox, P. M., Friedlingstein, P., Hugelius, G., & Westermann, S. (2017). An observation-based constraint on permafrost as a function of global warming. *Nature Climate Change, 7*(5), 340–344. https://doi.org/10.1038/nclimate3262

CIRNAC (Crown-Indigenous Relations and Northern Affairs Canada). (2019). *Canada's Arctic and northern policy framework*. Retrieved from https://www.rcaanc-cirnac.gc.ca/eng/1560523306861/1560523330587

Coote, M. L. (2016). Environmental decision-making in the Arctic Council: What is the role of Indigenous peoples? In L. Heininen, H. Exner-Pirot, & J. Plouffe (Eds.), *Arctic yearbook 2016* (pp. 42–56). Northern Research Forum.

Coote, M. L. (2023). The balance between critical thinking and paradigm thinking in the Arctic: Scientific cooperation across theoretical divides. *Polar Record, 59*, e5. https://doi.org/10.1017/S0032247422000377

Dewey, J. (1929). *The quest for certainty*. George Allen & Unwin.

DFAIT (Department of Foreign Affairs and International Trade). (2000). *The northern dimension of Canada's foreign policy*. Retrieved from https://library.arcticportal.org/1255/1/The_Northern_Dimension_Canada.pdf

Dittmer, J., Moisio, S., Ingram, A., & Dodds, K. (2011). Have you heard the one about the disappearing ice? Recasting Arctic geopolitics. *Political Geography, 30*(4), 202–214. https://doi.org/10.1016/j.polgeo.2011.04.002

Dodds, K., & Sörlin, S. (2022). *Ice humanities: Living, working and thinking in a melting world*. Manchester University Press.

Dodds, K., Kuus, M., & Sharp, J. (2013). *The Ashgate research companion to critical geopolitics*. Routledge.

Doering, N. N., Dudeck, S., Elverum, S., Fisher, C., Henriksen, J. E., Herrmann, T. M., Kramvig, B., Laptander, R., Milton, J., Omma, E. M., Saxinger, G., Scheepstra, A. J. M., & Wilson, K. (2022). Improving the relationships between Indigenous rights holders and researchers in the Arctic: An invitation for change in funding and collaboration. *Environmental Research Letters, 17*(6), 065014. https://doi.org/10.1088/1748-9326/ac72b5

Duveau, S. (2021). Frozen data? Polar research and fieldwork in a pandemic era. *Polar Record, 57*, e34. https://doi.org/10.1017/S0032247421000541

Engels, A. (2005). The science–policy interface. *The Integrated Assessment Journal (IAJ), 5*(1), 7–26.

Entman, R. (1993). Framing: Toward clarification of a fractured paradigm. *Journal of Communication, 43*(4), 51–58. https://doi.org/10.1111/j.1460-2466.1993.tb01304.x

EPPR (Emergency Prevention, Preparedness and Response). (2010). *Strategic plan of action: Emergency Prevention, Preparedness and Response Working Group (EPPR)*. Retrieved from http://hdl.handle.net/11374/1004

EPPR (Emergency Prevention, Preparedness and Response Working Group). (2024). *About*. Retrieved January, 29, 2026, from https://eppr.org/about/

Esposito, E. (2007). *Die Fiktion der Wahrscheinlichen Realität*. Suhrkamp.

Esposito, E. (2017). Beyond the promise of security: Uncertainty as resource. *Telos, 170*, 89–107.

EU-PolarNet 2 Consortium. (2022). *Directory of European Polar Research funding programmes*. Retrieved from https://eu-polarnet.eu/wp-content/uploads/2022/09/EU-PolarNet-2-Directory-EU-Polar-Research-Funding.pdf

Ford, J. D., Stephenson, E., Cunsolo Willox, A., Edge, V., Farahbakhsh, K., Furgal, C., Harper, S., Chatwood, S., Mauro, I., Pearce, T., Austin, S., Bunce, A., Bussalleu, A., Diaz, J., Finner, K., Gordon, A., Huet, C., Kitching, K., Lardeau, M.-P., … & Sherman, M. (2016). Community-based adaptation research in the Canadian Arctic. *Wiley Interdisciplinary Reviews: Climate Change, 7*(2), 175–191. https://doi.org/10.1002/wcc.376

Freistein, K., & Gadinger, F. (2022). Deconstructing the "hollow man": Visual narrative analysis and world politics. In P. Dawson & M. Mäkelä (Eds.), *Routledge companion to narrative theory* (pp. 512–526). Routledge.

Fujimura, J. (1992). Crafting science: Standardized packages, boundary objects, and "translation". In A. Pickering (Ed.), *Science as practice and culture* (pp. 168–212). University of Chicago Press.

Gehrke, C., & Wehrmann, D. (2024, October 18). Evidence-based policymaking in pessimistic times: Germany's new Arctic guidelines. *High North News*. Retrieved from https://www.highnorthnews.com/en/evidence-based-policymaking-pessimistic-times-germanys-new-arctic-guidelines
German Arctic Office. (2025). *Arctic Dialogue*. Retrieved January 29, 2025, from https://www.arctic-office.de/en/policy-advice/arctic-dialogue/
Gieryn, T. F. (1983). Boundary-work and the demarcation of science from non-science: Strains and interests in professional ideologies of scientists. *American Sociological Review, 48*(6), 781–795.
Gieryn, T. F. (1998). *Cultural boundaries of science: Credibility on the line*. University of Chicago Press.
Global Affairs Canada. (2010). *Statement of Canada's Arctic foreign policy: Exercising sovereignty and promoting Canada's Northern Strategy abroad*. Retrieved from https://publications.gc.ca/site/eng/9.833064/publication.html
Goffman, E. (1974). *Frame analysis: An essay on the organization of experience*. Harvard University Press.
Government of Canada. (2021, July 3). *Networks of Centres of Excellence*. Retrieved January 29, 2026, from https://science.gc.ca/site/science/en/interagency-research-funding/co-operative-funding-programs/networks-centres-excellence
Government of Canada. (2025a). *About Polar Knowledge Canada*. Retrieved January 29, 2026, from https://www.canada.ca/en/polar-knowledge/behindthescenes.html
Government of Canada. (2025b). *Provinces and territories*. Retrieved January 29, 2026, from https://www.canada.ca/en/intergovernmental-affairs/services/provinces-territories.html
Grinnell, F. (1992). *The scientific attitude*. Guilford Press.
Gruber, S., Hayley, J., Karunaratne, K., King, J., MacLean, T., Marshall, S., & Moore, D. (2023). Considerations toward a vision and strategy for permafrost knowledge in Canada. *Arctic Science, 9*(4), iv–viii. https://doi.org/10.1139/as-2023-0016
Gustafsson, K. M., & Lidskog, R. (2018). Boundary organizations and environmental governance: Performance, institutional design, and conceptual development. *Climate Risk Management, 19*, 1–11. https://doi.org/10.1016/j.crm.2017.11.001
Guston, D. H. (1999). Stabilizing the boundary between US politics and science: The role of the Office of Technology Transfer as a boundary organization. *Social Studies of Science, 29*(1), 87–111. https://doi.org/10.1177/030631299029001004
Guston, D. H. (2000). *Between politics and science: Assuring the integrity and productivity of research*. Cambridge University Press.
Guston, D. H. (2001). Boundary organizations in environmental policy and science: An introduction. *Science, Technology & Human Values, 26*(4), 399–408. https://doi.org/10.1177/016224390102600401
Haas, P. M. (1992). Introduction: Epistemic communities and international policy coordination. *International Organization, 46*(1), 1–35.
Haas, P. M. (2013). Epistemic communities. In J. Krieger (Ed.), *The Oxford companion to comparative politics* (Vol. 1, pp. 351–359). Oxford University Press.
Haavisto, P. (2001). *Review of the Arctic Council structures. Consultant's study.* The Finnish Institute of International Affairs. Retrieved from https://oaarchive.arctic-council.org/server/api/core/bitstreams/7ea0e112-179e-4933-9433-7efb9cac6c4b/content
Hagedorn, G., Loew, T., Seneviratne, S. I., Lucht, W., Beck, M.-L., Hesse, J., Knutti, R., Quaschning, V., Schleimer, J.-H., Mattauch, L., Breyer, C., Hübener, H., Kirchengast, G., Chodura, A., Clausen, J., Creutzig, F., Darbi, M., Daub, C.-H., Ekardt, F., … & Zens, J. (2019). The concerns of the young protesters are justified: A statement by scientists for future concerning the protests for more climate protection. *GAIA—Ecological Perspectives for Science and Society, 28*(2), 79–87. https://doi.org/10.14512/gaia.28.2.3
Hagg, W. (2022). *Glaciology and glacial geomorphology*. Springer.
Hall, P. A. (1993). Policy-paradigms, social learning and the state: The case of economic policymaking in Britain. *Comparative Politics, 25*(3), 275–296.

Hart, M. A. (2010). Indigenous worldviews, knowledge, and research: The development of an Indigenous research paradigm. *Journal of Indigenous Voices in Social Work, 1*(1), 1–16.
Heinemann, G., Braun, M., Brey, T., Damaske, D., Melles, M., Rhein, M., & Willmes, S. (Eds.). (2017). *Polarforschungsagenda 2030: Status und Perspektiven der deutschen Polarforschung*. Statusbericht des Deutschen Nationalkomitees SCAR/IASC der DFG. Retrieved from https://www.dfg.de/resource/blob/173024/polarforschungsagenda-2030.pdf
Helmholtz Gemeinschaft. (2006). *Geschäftsbericht 2006*. Retrieved from https://www.helmholtz.de/fileadmin/user_upload/publikationen/Geschaeftsberichte/Helmholtz_GB_2006.pdf
Helmholtz Gemeinschaft. (2026). *Program-oriented funding*. Retrieved January 29, 2026, from https://www.helmholtz.de/en/about-us/structure-and-governance/program-oriented-funding/
Hoppe, R. (1999). Policy analysis, science, and politics: From "speaking truth to power" to "making sense together". *Science and Public Policy, 26*(3), 201–210. https://doi.org/10.3152/147154399781782482
Hoppe, R. (2005). Rethinking the science-policy nexus: From knowledge utilization and science technology studies to types of boundary arrangements. *Poiesis & Praxis, 3*(3), 199–215. https://doi.org/10.1007/s10202-005-0074-0
Ibarguchi, G., Rajdev, V., & Murray, M. S. (2018). Are current research funding structures sufficient to address rapid Arctic change in a meaningful way? *Polar Research, 37*, 1540242. https://doi.org/10.1080/17518369.2018.1540242
INAC (Indian and Northern Affairs Canada). (2009). *Canada's Northern Strategy: Our North, our heritage, our future*. Retrieved from https://publications.gc.ca/collections/collection_2009/ainc-inac/R3-72-2008.pdf
IPCC (Intergovernmental Panel on Climate Change). (2019). *IPCC special report on the ocean and cryosphere in a changing climate*. Cambridge University Press.
Isberg, E. (2022). Frozen archives on the go: Ice cores and the temporalization of earth system science. In K. Dodds & S. Sörlin (Eds.), *Ice humanities: Living, thinking and working in a melting world* (pp. 266–283). Manchester University Press.
Jasanoff, S. (1987). Contested boundaries in policy-relevant science. *Social Studies of Science, 17*(2), 195–230.
Jasanoff, S. (2005). *Designs on nature: Science and democracy in Europe and the United States*. Princeton University Press.
Jasanoff, S. (2012). *Science and public reason*. Routledge.
Jasanoff, S. (2015). Future imperfect: Science, technology, and the imaginations of modernity. In S. Jasanoff & S.-H. Kim (Eds.), *Dreamscapes of modernity: Sociotechnical imaginaries and the fabrication of power* (pp. 1–33). University of Chicago Press.
Jasanoff, S., & Kim, S.-H. (Eds.). (2015). *Dreamscapes of modernity: Sociotechnical imaginaries and the fabrication of power*. University of Chicago Press.
Jungcurt, S. (2013). Taking boundary work seriously: Towards a systemic approach to the analysis of interactions between knowledge production and decision-making on sustainable development. In L. Meuleman (Ed.), *Transgovernance: Advancing sustainability governance* (pp. 255–273). Springer.
Kaldewey, D. (2018). The grand challenges discourse: Transforming identity work in science and science policy. *Minerva, 56*(2), 161–182. https://doi.org/10.1007/s11024-017-9332-2
Kallerud, E., Langfeld, L., & Søgnen, R. (2001). *RCN budgets, policy instruments and operations. Background report No 7 in the evaluation of the Research Council of Norway.* NIFU—Norsk Institutt for Studier av Forskning og Utdanning. Retrieved from https://www.regjeringen.no/globalassets/upload/kd/vedlegg/forskning/rapporter/2001-rcn-eval/2001-rcn-evaluation-background-report-no-7.pdf?id=2248236
Kampourakis, K., & McCain, K. (2019). *Uncertainty: How it makes science advance*. Oxford University Press.
Kass, G., Milner, A. M., & Dodds, K. (2022). The "borderlands" of the science–policy interface. *Geographical Journal, 188*(4), 591–599. https://doi.org/10.1111/geoj.12469

Katzenstein, P. J. (2022). Worldviews in world politics. In P. J. Katzenstein (Ed.), *Uncertainty and its discontents: Worldviews in world politics* (pp. 1–69). Cambridge University Press.
Kay, J., & King, M. (2020). *Radical uncertainty: Decision-making for an unknowable future*. Bridge Street Press.
Keskitalo, E. C. H. (2012). Setting the agenda on the Arctic: Whose policy frames the region? *The Brown Journal of World Affairs, 19*(1), 155–164.
Klerkx, L., & Leeuwis, C. (2008). Delegation of authority in research funding to networks: Experiences with a multiple goal boundary organization. *Science and Public Policy, 35*(3), 183–196. https://doi.org/10.3152/030234208X299053
Knecht, S., & Albert, M. (2022). The Arctic is what scientists make of it: Integrating geopolitics into informed decisionmaking. In P. A. Berkman, A. N. Vylegzhanin, O. A. Young, D. A. Balton, & O. R. Øvretveit (Eds.), *Building common interests in the Arctic Ocean with global inclusion* (Vol. 2, pp. 365–382). Springer.
Koselleck, R. (2004). *Futures past: On the semantics of historical time*. MIT Press.
Lackenbauer, W. P. (2020). *Canada's Northern Strategies: From Trudeau to Trudeau, 1970–2020*. Documents on Canadian Arctic Sovereignty and Security No. 17. Retrieved from https://arts.ucalgary.ca/sites/default/files/teams/25/17%20-%20Lackenbauer%20-%20Canadas%20Northern%20Strategies.pdf
Langer, M., Schneider von Deimling, T., Westermann, S., Rolph, R., Rutte, R., Antonova, S., Rachold, V., Schultz, M., Oehme, A., & Grosse, G. (2023). Thawing permafrost poses environmental threat to thousands of sites with legacy industrial contamination. *Nature Communications, 14*, 1721. https://doi.org/10.1038/s41467-023-37276-4
Latour, B. (1987). *Science in action: How to follow scientists and engineers through society*. Harvard University Press.
Latour, B. (2004). *Politics of nature: How to bring the sciences into democracy*. (C. Porter, Trans.). Harvard University Press.
Law, J. (2004). *After method: Mess in social science research*. Routledge.
Lepori, B. (2011). Coordination modes in public funding systems. *Research Policy, 40*(3), 355–367. https://doi.org/10.1016/j.respol.2010.10.016
Łuszczuk, M., Padrtova, B., & Szczerbowicz, W. (2020). Political dimension of Arctic research. *Oceanologia, 62*(4), 608–621. https://doi.org/10.1016/j.oceano.2020.03.008
Maasen, S., Kaiser, M., Reinhart, M., & Sutter, B. (Eds.). (2012). *Handbuch Wissenschaftssoziologie*. Springer.
Machen, R. (2018). Towards a critical politics of translation: (Re)producing hegemonic climate governance. *Environment and Planning E: Nature and Space, 1*(4), 494–515. https://doi.org/10.1177/2514848618785515
MacKenzie, D., & Millo, Y. (2003). Constructing a market, performing theory: The historical sociology of a financial derivatives exchange. *American Journal of Sociology, 109*(1), 107–145.
Mahony, M. (2013). Boundary spaces: Science, politics, and the epistemic geographies of climate change in Copenhagen, 2009. *Geoforum, 49*, 29–39. https://doi.org/10.1016/j.geoforum.2013.05.005
Matthes, J., & Kohring, M. (2008). The content analysis of media frames: Toward improving reliability and validity. *Journal of Communication, 58*(2), 258–279. https://doi.org/10.1111/j.1460-2466.2008.00384.x
Mayer, C., Hagg, W., Weber, M., & Lambrecht, A. (2021). *Zukunft ohne Eis: Zweiter Bayerischer Gletscherbericht: Klimawandel in den Alpen*. Bayerische Akademie der Wissenschaften, Bayerisches Staatsministerium für Umwelt und Verbraucherschutz. Retrieved from https://badw.de/fileadmin/user_upload/Files/BADW/Neuigkeiten/2021/04-21/Bayerischer_Gletscherbericht_2021_bf_low-DS_n.pdf
Morison, J., Aagaard, K., & Steele, M. (1998). *Report on study of the Arctic Change Workshop held November 10–12, 1997*. Polar Science Center, University of Washington. Retrieved from https://www.arcus.org/files/publication/24591/search_1997_arctic_change_report.pdf

Morlighem, M., Williams, C. N., Rignot, E., An, L., Arndt, J. E., Bamber, J. L., Catania, G., Chauché, N., Dowdeswell, J. A., Dorschel, B., Fenty, I., Hogan, K., Howat, I., Hubbard, A., Jakobsson, M., Jordan, T. M., Kjeldsen, K. K., Millan, R., Mayer, L., … & Zinglersen, K. B. (2017). BedMachine v3: Complete bed topography and ocean bathymetry mapping of Greenland from multibeam echo sounding combined with mass conservation. *Geophysical Research Letters*, *44*(21), 11051–11061. https://doi.org/10.1002/2017GL074954.

Nash, M. (2021). National Antarctic program responses to fieldwork sexual harassment. *Antarctic Science, 33*(5), 560–571. https://doi.org/10.1017/S0954102021000432

Norwegian Ministry of Foreign Affairs. (2011). *The High North: Vision and strategies*. Retrieved from https://www.regjeringen.no/globalassets/upload/ud/vedlegg/nordomradene/ud_nordomrodene_en_web.pdf

Nowotny, H., Scott, P., & Gibbons, M. (2001). *Re-thinking science: Knowledge and the public in the age of uncertainty*. Polity.

Nuth, C., Kohler, J., König, M., von Deschwanden, A., Hagen, J. O., Kääb, A., Moholdt, G., & Pettersson, R. (2013). Decadal changes from a multi-temporal glacier inventory of Svalbard. *The Cryosphere, 7*(5), 1603–1621. https://doi.org/10.5194/tc-7-1603-2013

O'Reilly, J. (2017). *The technocratic Antarctic: An ethnography of scientific expertise and environmental governance*. Cornell University Press.

Osipov, I. A., Radford, G., Aksnes, D. W., Kullerud, L., & Hirshberg, D. (2016). *International Arctic Research: Analyzing global funding trends. A pilot report*. https://doi.org/10.6084/m9.figshare.3811224.

Osipov, I. A., Radford, G., Aksnes, D. W., Kullerud, L., & Hirshberg, D. (2017). *International Arctic Research: Analyzing global funding trends. A pilot report (2017 update)*. https://doi.org/10.6084/m9.figshare.4829455.

Østhagen, A. (2021). Norway's Arctic policy: Still high north, low tension? *The Polar Journal, 11*(1), 75–94. https://doi.org/10.1080/2154896X.2021.1911043

Østhagen, A. (2024). The myths of Svalbard geopolitics: An Arctic case study. *Marine Policy, 167*, 106183. https://doi.org/10.1016/j.marpol.2024.106183

PAME (Protection of the Arctic Marine Environment). (2003). *PAME report to Senior Arctic Officials. Attachment I.* Reykjavík, Iceland. Retrieved April 9–10, 2003, from http://hdl.handle.net/11374/568

PAME (Protection of the Arctic Marine Environment). (2024). *About*. Retrieved January 29, 2026, from https://pame.is/about/

POLAR (Polar Knowledge Canada). (2020a). *Strategic plan*. Retrieved from https://www.canada.ca/content/dam/polar-polaire/documents/pdf/strategic-plan/strategic-plan-EN.pdf

POLAR (Polar Knowledge Canada). (2020b). *Science and technology framework 2020–2025*. Retrieved from https://www.canada.ca/content/dam/polar-polaire/documents/pdf/science-technology-framework/science-technology-framework-2020-2025-EN.pdf

Powell, R. C. (2017). *Studying Arctic fields: Cultures, practices, and environmental sciences*. McGill–Queen's University Press.

RCN (Research Council of Norway). (2010). *Norwegian polar research: Policy for Norwegian polar research 2010–13*. Retrieved from https://www.forskningsradet.no/siteassets/publikasjoner/1253955848791.pdf

RCN (Research Council of Norway). (2011). *Research strategy for the Arctic and Northern areas: Revision 1 2011–2016*. Retrieved from https://www.forskningsradet.no/siteassets/publikasjoner/1253970528732.pdf

RCN (Research Council of Norway). (2013). *Norwegian polar research: Research policy 2014–2023*. Retrieved from https://www.forskningsradet.no/siteassets/publikasjoner/1253993655002.pdf

RCN (Research Council of Norway). (2015). *Work programme from 2015: Polar Research Programme—POLARPROG*. Retrieved from https://www.forskningsradet.no/siteassets/publikasjoner/1254021856251.pdf

RCN (Research Council of Norway). (2018). *Work programme from 2018. Polar Research Programme—POLARPROG*. Retrieved from https://www.forskningsradet.no/contentassets/f7a4355bbb7a4aaabab5ee7df1bd34bb/polarprog-work-programme.pdf

RCN (Research Council of Norway). (2019, April 10). *The Research Council of Norway through 30 years*. Retrieved January 29, 2026, from https://www.forskningsradet.no/en/about/what-we-do/

RCN (Research Council of Norway). (2025a). *The centres for research-based innovation scheme*. Retrieved January 29, 2026, from https://www.forskningsradet.no/en/financing/what/sfi/

RCN (Research Council of Norway). (2025b). *Norwegian Centres of Excellence scheme*. Retrieved January 29, 2026, from https://www.forskningsradet.no/en/financing/what/sff/

RCN (Research Council of Norway). (2025c, October 15). *About Svalbard Science Forum (SSF)*. Retrieved January 29, 2026, from https://www.forskningsradet.no/en/svalbard-science-forum/about-ssf/

Rosales, J., & Chapman, J. L. (2015). Perceptions of obvious and disruptive climate change: Community-based risk assessment for two native villages in Alaska. *Climate, 3*(4), 812–832. https://doi.org/10.3390/cli3040812

Rottem, S. V. (2020). Improving the work of the Arctic Council: Challenges and recommendations. *Polar Record, 57*, e49.

Rottem, S. V., Soltvedt, I. F., & Prip, C. (2020). Arktisk råd i spennet mellom forskning, forvaltning og politikk. *Internasjonal Politikk, 78*(3), 284–310. https://doi.org/10.23865/intpol.v78.1504

Ruffini, P.-B. (2020). Conceptualizing science diplomacy in the practitioner-driven literature: A critical review. *Humanities & Social Sciences Communications, 7*, 124. https://doi.org/10.1057/s41599-020-00609-5

Salewski, C., Krause, R., & Angele, E. (2020). Das Alfred-Wegener-Institut in der Geschichte der Polarforschung. Einführung und Chronik. *Reports on Polar and Marine Research, 742*. https://doi.org/10.2312/BzPM_0742_2020

Sanne, D. (2008). *Perpetual repercussion, Svalbard Global Seed Vault*. Retrieved January 29, 2026, from https://www.dyvekesanne.com/www.dyvekesanne.com/Global.html

Schattschneider, E. E. (1960). *The semi-sovereign people: A realist's view of democracy in America*. Holt, Rinehart and Winston.

Schiermeier, Q. (2017). Time capsule buried to preserve science for the ages. *Nature*. https://doi.org/10.1038/nature.2017.22657

Schreier, M. (2012). *Qualitative analysis in practice*. Sage.

Schütz, A. (1959). Tiresias, or our knowledge of future events. *Social Research, 26*(1), 71–89.

Scoones, I., & Stirling, A. (2020). *The politics of uncertainty: Challenges of transformation*. Routledge.

Scrivener, D. (1998). From Ottawa to Iqaluit: Towards sustainable Arctic co-operation? *Environmental Politics, 7*(4), 136–141.

SDWG (Sustainable Development Working Group). (2014). *Sustainable Development Working Group—Progress Report to Senior Arctic Officials for the Senior Arctic Officials Meeting Number 2*. 26–27 March, 2014, Yellowknife, NWT. Retrieved from http://hdl.handle.net/11374/1298

SDWG (Sustainable Development Working Group). (2024). *About us*. Retrieved January 29, 2026, from https://sdwg.org/about/

Sellheim, N., & Menezes, D. R. (2022). *Non-state actors in the Arctic region*. Springer.

Serreze, M. C. (2018). *Brave New Arctic: The untold story of the melting North*. Princeton University Press.

Shugar, D. H., Clague, J. J., Best, J. L., Schoof, C., Willis, M. J., Copland, L., & Roe, G. H. (2017). River piracy and drainage basin reorganization led by climate-driven glacier retreat. *Nature Geoscience, 10*, 370–375. https://doi.org/10.1038/ngeo2932

Simons, A., & Schniedermann, A. (2021). The neglected politics behind evidence-based policy: Shedding light on instrument constituency dynamics. *Policy & Politics, 49*(4), 513–529. https://doi.org/10.1332/030557321X16225469993170

Skagestad, G. (2010). *The "High North": An elastic concept in Norwegian Arctic policy*. Fridtjof Nansen Institute, FNI Report 10. https://www.fni.no/getfile.php/133828-1756148688/1.%20FNI%20-%20nettside/Import%20fra%20fni.no/Filer/Publikasjoner/FNI-R1010.pdf

Skoie, H. (2000). Diversity and identity: The merger of five research councils in Norway. *Science and Public Policy, 27*(2), 83–96. https://doi.org/10.3152/147154300781782084

Skupien, S., & Rüffin, N. (2020). The geography of research funding: Semantics and beyond. *Journal of Studies in International Education, 24*(1), 24–38. https://doi.org/10.1177/1028315319889896

Snow, D. A., & Benford, R. D. (1988). Ideology, frame resonance, and participant mobilization. *International Social Movement Research, 1*, 197–217.

Spence, J. (2017). Is a melting Arctic making the Arctic Council too cool? Exploring the limits to the effectiveness of a boundary organization. *Review of Policy Research, 34*(6), 790–811. https://doi.org/10.1111/ropr.12257

Star, S. L., & Griesemer, J. R. (1989). Institutional ecology, "translations", and boundary objects: Amateurs and professionals in Berkeley's Museum of Vertebrate Zoology, 1907–39. *Social Studies of Science, 19*(3), 387–420. https://doi.org/10.1177/030631289019003001

Stefansson, V. (1945). *The Arctic in fact and fable. Headline Series, 51.* Foreign Policy Association. Retrieved January, 29, 2026, from https://archive.org/details/sim_headline-series_march-april-1945_51/page/4/mode/2up

Steffensen, J. P. (2012). Willi Dansgaard. *Polar Record, 48*(2), 206–207. https://doi.org/10.1017/S0032247411000155

Steinberg, P. E., Tasch, J., & Gerhardt, H. (2018). *Contesting the Arctic: Politics and imaginaries in the circumpolar north*. I.B. Tauris.

Stokke, O. S., & Hønneland, G. (2007). *International cooperation and Arctic governance: Regime effectiveness and northern regime building*. Routledge.

Straßheim, H. (2021). Time, sustainability, and public policy. In K. H. Goetz (Ed.), *The Oxford handbook of time and politics* (pp. 285–306). Oxford University Press.

Straßheim, H. (2024). A world of evidence: The global spread and silent politics of evidence cultures. *Review of Policy and Society, 43*(4), 414–431. https://doi.org/10.1093/polsoc/puae029

Tallberg, J., Sommerer, T., Squatrito, T., & Jönsson, C. (2013). Explaining transnational access to international organizations: Theories and hypotheses. In J. Tallberg, T. Sommerer, T. Squatrito, & C. Jönsson (Eds.), *opening up Up of international organizations: Transnational access in global governance* (pp. 22–52). Cambridge University Press.

Tennberg, M. (2000). *Arctic environmental cooperation: A study in governmentality*. Routledge.

The IMBIE Team. (2018). Mass balance of the Antarctic ice sheet from 1992 to 2017. *Nature, 558*, 219–222. https://doi.org/10.1038/s41586-018-0179-y

The Royal Society. (2010). *New frontiers in science diplomacy: Navigating the changing balance of power.* Retrieved from https://royalsociety.org/-/media/policy/publications/2010/4294969468.pdf

Tuathail, G. O., & Agnew, J. (1992). Geopolitics and discourse: Practical geopolitical reasoning in American foreign policy. *Political Geography, 11*(2), 190–204. https://doi.org/10.1016/0962-6298(92)90048-X

UNFCCC (United Nations Framework Convention on Climate Change). (2022). *Sharm el-Sheikh Implementation Plan. Decision -/CP.27.* Retrieved from https://unfccc.int/documents/624444

Väätänen, V. (2022). The construction, solidification and political implications of geographical scientific facts: A perspective on the changing Arctic region. *Geoforum, 128*, 21–32. https://doi.org/10.1016/j.geoforum.2021.11.019

van der Meulen, B. (1998). Science policies as principal–agent games: Institutionalization and path dependency in the relation between government and science. *Research Policy, 27*(4), 397–414. https://doi.org/10.1016/S0048-7333(98)00049-3

van Soest, M. (2023). An Arctic expedition: A supposedly useful thing I'll never do again. *Polar Record, 42*. https://doi.org/10.33265/polar.v42.9070

Westerveld, L., Kurvits, T., Schoolmeester, T., Mulelid, O. B., Eckhoff, T. S., Overduin, P. P., Fritz, M., Lantuit, H., Alfthan, B., Sinisalo, A., Miesner, F., Viitanen, L.-K., & The NUNATARYUK

Consortium. (2023). *Arctic permafrost atlas*. GRID-Arendal. Retrieved from https://www.grida.no/publications/998

Wildavsky, A. (2007). *Speaking truth to power: The art and craft of policy analysis (12th printing)*. Transaction Publishers.

Wilson Rowe, E. (2018). *Arctic governance: Power in cross-border cooperation*. Manchester University Press.

Wissenschaftlicher Begleitkreis Mare:N "Forschungsagenda Polarregionen im Wandel". (2021). *Forschungsagenda Polarregionen im Wandel. Konzeptpapier des Mare:N-Begleitkreises*. Retrieved from https://www.fona.de/medien/pdf/Forschungsagenda_Polarregionen_Konzeptpapier_des_MAREN_Begleitkreises_Mai_2021.pdf

Wolff, L. (2021). The past shall not begin: Frozen seeds, extended presents and the politics of reversibility. *Security Dialogue, 52*(1), 79–95. https://doi.org/10.1177/0967010620912961

Wood-Donnelly, C. (2019). *Performing Arctic sovereignty: Policy and visual narratives*. Routledge.

Wood-Donnelly, C., & Bartels, M. P. (2022). Science diplomacy in the Arctic: Contributions of the USGS to policy discourse and impact on governance. *Polar Record, 58*, e16. https://doi.org/10.1017/S0032247422000134

Young, O. (1998). *Creating regimes: Arctic accords and international governance*. Cornell University Press.

Yua, E., Raymond-Yakoubian, J., Daniel, R. A., & Behe, C. (2022). A framework for co-production of knowledge in the context of Arctic research. *Ecology and Society, 27*(1), 34. https://doi.org/10.5751/ES-12960-270134